Mathias Bach

Near Infrared Laser Sensor System for In-Line Detection of Conversion in UV-Cured polymer Coatings

Near Infrared Laser Sensor System for In-Line Detection of Conversion in UV-Cured polymer Coatings

by
Mathias Bach

Dissertation, Karlsruher Institut für Technologie (KIT)
Fakultät für Elektrotechnik und Informationstechnik, 2012

Impressum

Karlsruher Institut für Technologie (KIT)
KIT Scientific Publishing
Straße am Forum 2
D-76131 Karlsruhe
www.ksp.kit.edu

KIT – Universität des Landes Baden-Württemberg und nationales
Forschungszentrum in der Helmholtz-Gemeinschaft

KIT Scientific Publishing 2012
Print on Demand

ISBN 978-3-86644-839-1

Near Infrared Laser Sensor System for In-Line Detection of Conversion in UV-Cured Polymer Coatings

Zur Erlangung des akademischen Grades eines
DOKTOR-INGENIEURS

an der Fakultät für
Elektrotechnik und Informationstechnik
des Karlsruher Instituts für Technologie KIT
genehmigte

DISSERTATION
von
Dipl.-Geophys. Mathias Bach
geb. in Mutlangen

Hauptreferent: Prof. Dr.-Ing. Wolfgang Heering
Korreferent: Prof. Dr. rer. nat. Wilhelm Stork

Tag der mündlichen Prüfung: 24. April 2012

Abstract | Kurzbeschreibung

Abstract of the present work

The present work describes a method for the determination of the conversion of C-double bonds in acrylic coatings by radical photopolymerization that is suitable for an in situ monitoring during the coating process. Acrylate based coatings are increasingly used in many kinds of industrial coating processes, due to advantages of the solvent free application process.

The applied method is based on the $1620\,nm$ overtone absorption of the acrylate C-double bond end group. To overcome the intensity problems of common near infrared spectrophotometers, the diffuse transflection only at the correlated discrete wavelength of $1620\,nm$ and at two reference wavelengths is measured. A sophisticated sensor set-up including an optical spatial filter and lock-in amplified signal processing is used to achieve a sufficient signal to noise ratio for the applicability in the final prototype.

The capability of the sensor to discriminate between polymerized and unpolymerized coatings on metal substrates down to a coating thickness of less than $16\,\mu m$ is demonstrated and proved by reference measurements with ATR-FTIR and NIR equipment. Furthermore, the information on polymerization from larger depths in TiO_2 pigmented coatings has been investigated with a customized laboratory set-up. The results can be analytically modelled in analogy to the Lambert-Beer's law, resulting in a detection limit of a maximum pigment concentration of $15\,wt\%$ with $100\,\mu m$ coating thickness. The presented final sensor design is suitable to be used in an industrial production environment for example in screen printing applications as a monitoring and quality control tool.

Kurzbeschreibung der vorliegenden Arbeit

Die vorliegende Arbeit beschreibt ein Verfahren zur Bestimmung des Umsetzungsgrades von C-Doppelbindungen in UV härtenden Acrylatbeschichtungen. Das entwickelte Messprinzip ist geeignet für eine in-situ Überwachung während der Beschichtung in industriellen Druck- und Lackierprozessen. UV härtende Acrylate werden immer häufiger in vielen Arten von industriellen Beschichtungsverfahren eingesetzt. Eine hohe Oberflächenhärte und ein sehr guter Korrosionsschutz zeichnen diese Beschichtungen aus.

Weitere Vorteile der UV härtenden Acrylate bestehen in der Lösemittelfreiheit und kurzen Polymerisationszeiten, welche eine sofortige Weiterverarbeitung ermöglichen. Der entwickelte diskrete Lasersensor nutzt die $1620\,nm$ Oberton-Schwingung der Acrylatendgruppe, deren C-Doppelbindungen bei der Polymerisation aufgebrochen werden. Die diffuse Transflektion der die Beschichtung passierenden Laserstrahlung bei $1620\,nm$ wird mit einer InGaAs-Photodiode detektiert. Zusätzlich sind weitere, von der Polymerisation unabhängige Lasersignale notwendig, um eine Normalisierung der Signale zu ge- währleisten.

Ein Sensorkopf mit optischen Raumfilter und Lock-In Verstärker wurde entwickelt, um ein ausreichendes Signal-Rausch-Verhältnis für die Anwendung im endgültigen Prototyp zu erreichen. Es wird eine Unterscheidung zwischen polymerisierten und nicht polymerisierten Schichten auf metallischen Substraten bis zu einer Schichtdicke von weniger als $16\,\mu m$ demonstriert und dieses Ergebnis durch Vergleichsmessungen mit bekannten ATR-FTIR und NIR Techniken verifiziert. Darüber hinaus wird die Polymerisation von TiO_2 pigmentierten Lacken bis Schichtdicken von mehreren $100\,\mu m$ untersucht. Die Ergebnisse lassen sich in Analogie zu dem Lambert-Beer-Gesetz modellieren. Die Nachweisgrenze des Umsetzungsgrades bei einer maximalen Pigmentkonzentration von $15\,wt\%$ wird nachweislich bei einer Schichtdicke von $100\,\mu m$ erreicht. Der Prototyp wurde bereits für die Überwachung des Umsetzungsgrades bei einem Siebdruckprozess eingesetzt und erste Ergebnisse werden vorgestellt.

Relating publications, patents, presentations and posters by the author:

- 10. Karlsruher Arbeitsgespräche: Produktionsforschung 2010; Exhibition of the sensor and poster session

- M. Bach, S. Pieke, C. Kaiser, R. Kling, W. Heering, Europtrode X, Prague, Poster,2010

- M. Bach, O. Treichel, E. Feuerbacher, S. Pieke, R. Kling, W. Heering; Discrete NIR Laser Spectroscopy for Inline Monitoring of Acrylic UV Curing Processes; European Coatings Conference, Nürnberg 2011, Talk, 2011

- M. Bach, O. Treichel, E. Feuerbacher, W. Heering; "Cured by light, controlled by light", *European Coatings Journal*, Technical paper, 10, pp. 44-48, 2011

- Bach, M., Feuerbacher, E., Treichel, O., Heering, W., Pieke, S., Drexler, J., Eberhardt, M., Fornezza, D., "Optisches Sensorsystem für die Online-Messung der Vernetzung von Farben und Lacken in Druck- und Lackierprozessen (CuringOnlineSensor)", invention disclosure, 2011

- Bach, M. and Heering, W., NIR-laser based detection of the photopolymerization state of acrylate coatings suitable for in situ measurements. Journal of Applied Polymer Science, 124: 3494-3500. doi: 10.1002/app.35462, 2012

Contents

Chapter 1

Introduction

1.1 Objectives of this work

The objective of this thesis is the introduction of a new sensor technique for the detection of the conversion grade of photo-curable resins and coatings. The developed sensor system uses a new discrete laser-spectroscopic method, which provides significant improvements concerning the detection principle and the applicability.

UV-coating formulations do often contain acrylates, which are responsible for the polymerization of the coating. The acrylic C-double bond is cracked by the photoinitiator radicals, which are activated by the UV radiation. The opened C-bonds form a polymer network by linking multiple polymer fragments to a solid polymer chain. The main principle of the presented sensor is based up on the radiation absorbance attribute of the acrylate end group. This absorbance depends on the amount of cracked, respectively intact C-double bonds.

Spectroscopic means can measure the infrared absorbance due to oscillating transitions of molecular ligands. The more C-double bonds are cracked the less radiation at a distinct wavelength will be absorbed. These absorption spectra provide the fundamentals of the later presented discrete laser measurement of the acrylic conversion.

Today's spectrometers acquire the molecular absorption spectrum with the use of a broadband light source, for example a halogen incandescent lamp, and by dispersive elements like gratings or prisms. Currently there are available, only a few and expensive sensor systems used for an on-line monitoring of the production processes [1, 2], next to the laboratory based spectroradiometers for quality control.

In contrast to that, the here presented sensor system applies discrete lasers at distinct absorption wavelengths as radiation sources, thereby avoiding expensive diffractive and optical elements for selecting the appropriate wavelength. Furthermore, the sensor system is designed for a continuous, contact free and high speed measurement of the conversion grade in photocured polymer coatings.

The main driver for this approach originates from the need of a cost effective, reliable and fast monitoring tool in the UV-coating industry. Examples are the product safety of coatings on food packaging or coatings on security relevant parts in the automotive industry. Retracing and continuous tracking of each individual product is already most important in these sectors. Thus the introduction of such a sensor system, capable of tracking and monitoring of coating conversion, is highly demanded.

The presented work covers the complete research and development process: Understanding the challenges of measuring the chemical conversion in photo-curing systems, analysing the most effective detection mechanisms, designing of the laboratory demonstrator and finally the presentation of a complete sensor system, ready to be installed in the previous mentioned UV-curing applications. Main aspects within this research are the verification of the sensor signal with regard to the grade of conversion and the detection reliability and suitability of the sensor system on different coating formulations.

This work is based on the high risk research project "'CuringOnlineSensor"' (02PU2511) supported by the "'Bundesministerium für Bildung und Forschung"' (BMBF).

1.2   The transflection concept

As it was explained in the preliminary objectives, the measurement of the polymerization of coatings within thicknesses of a few micrometers up to some hundred micrometers is currently performed in laboratories and not suitable for any kind of industrial application. Therefore a new approach for the acquisition of conversion information from the coating surface, the inner part of the coating and from the substrate coating interface at the same time has to be developed.

Taking a closer look on the above mentioned conditions for the measurement a thorough understanding of two main optical principles, transmission and reflection of light and its interaction with the acrylic coating will

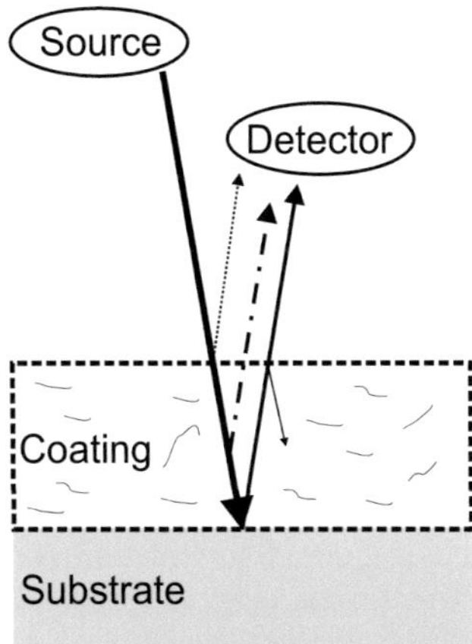

Figure 1.1: Illustration of the transflection definition used in this work. The incident light might be (a) reflected on the coating surface $(\cdots)$; (b) reflected from the coating volume $(\cdot - \cdot)$; (c) reflected from the substrate coating interface $(-)$; (c1) total reflection at the coating volume - air interface $(\leftarrow)$. The intensity of the laser light, which is transflected from the coating volume and the substrate is partially absorbed by the molecular oscillation of the acrylate, depending on the amount of cracked C-double bonds and thus depending on the conversion of the coating.

be crucial for the successes of such a sensor device. During the laboratory research and development of the here presented sensor principle, a more advanced definition for the interaction between reflection and transmission on the coatings has been established:

The sensor principle is defined as transflection of the used laser light on the coated samples. This means some part of the laser light is transmitted into the coating and finally reflected from the coating substrate interface, some other part is already reflected from the air-coating interface or reflected from the coating volume itself. The intensity of the laser light, which is transflected from the coating volume and the substrate is partially absorbed by the molecular oscillation of the acrylate, depending on the amount of cracked C-double bonds and thus depending on the conversion of the coating. Figure 1.1 illustrates this transflection principle measuring the conversion of the polymer.

1.3   The UV-coating industry

Since the early 20th century UV-radiation is used for irradiation of various kinds of materials and radiation induced processes [3]. The electron beam irradiation was one of the first radiation technologies used for producing wood flooring and simple composite substrates. In the 1960's the chemical industry developed the first photosensitive formulations, which are known today as photo-curing polymers. The key drivers for the increasing use of UV-curing processes are the more and more restricted emission of volatile organic compounds (VOCs) in the latest years and the demanded high performance of the products regarding scratch and chemical resistance. The emission regulation of VOCs was again reinforced by the European Commission in 2004 [4]. This regulation introduced a two step limit for volatile compounds of inks, coatings and varnishes. The VOC limit in primers for example has been set to 450 g/l in 2007 and was further reduced to 350 g/l in 2010. Compared to the limits of 1999 this is a reduction by about 70%. Due to this progress in health safety and environment protection, UV-curable, solvent free formulations are increasingly demanded by the market, especially by the automotive and printing industry.

1.3.1   Markets and applications of the UV-curing technology

This increasing demand for radiation curable coatings can be seen in Figure 1.2. The Figure shows the revenue and growth rate forecast for the UV-printing industry from 2002 to 2012, as predicted by Frost and Sullivan in 2006.

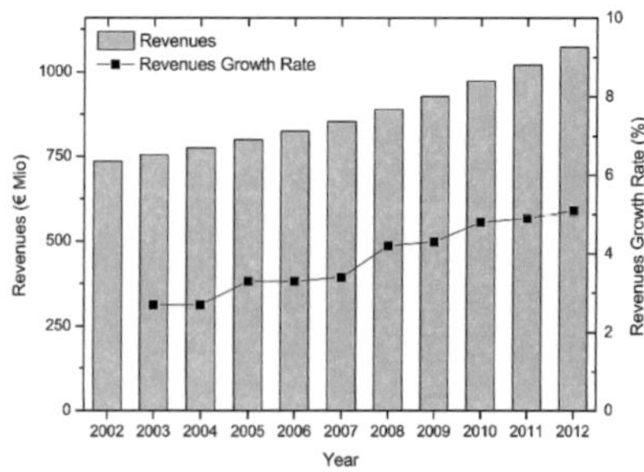

Figure 1.2: Radiation curable coatings market in Europe. A revenue and growth rate forecast from 2002 - 2012 in 2006 [5].

For 2012 an annual growth of about 5% and revenue of over € 1,1 billion is predicted for the European market. This revenue can be subdivided in three main curing techniques as shown in Figure 1.3. UV-VIS irradiation applications are clearly leading with a share of over 98%. Minor important in terms of revenues are electron beam and cationic applications. The almost exclusive UV-VIS application in the industry is explained by the effective and economic possibility of producing this part of the spectrum from 200 nm to 400 nm.

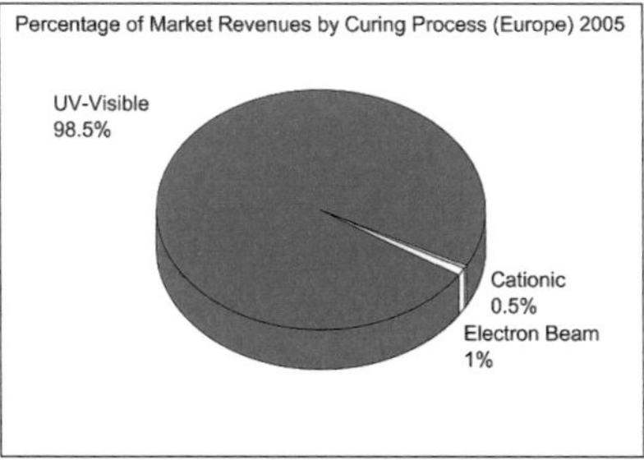

Figure 1.3: Market share of the mainly used photoinitiated curing techniques in 2005. The UV-VIS free radical photopolymerization is clearly leading with 98,5% of the total revenue achieved in 2005 [5].

An almost similar situation can be observed as looking on to the revenue shares of the types of resins used in the market. Acrylate resins, which are the major targets of this work, have the main share of 80%. Polyester resins are following with about 20% (see Figure 1.4). Acrylate resins do have significant advantages like higher reactivity, better gloss and less irritation compared to polyester resins.

All above mentioned important issues for the radiation curable coatings market lead to the conclusion that the acrylate resins in combination with UV-VIS radiation sources are the main driver for future developments in this business. Therefore it is obvious to concentrate the research affords onto developing tools for a better quality control and higher production output on this main sector of acrylates.

1.3.2 Market shares of the UV-curing technology

Like in the conventional coating and printing industry the application areas of UV-cured coatings are as various. There are five main areas, in which the

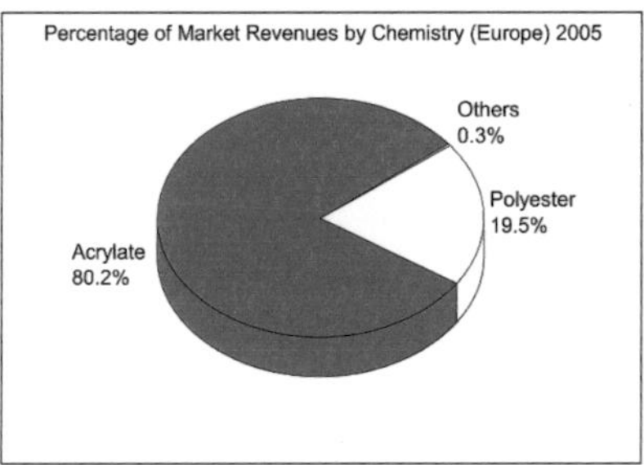

Figure 1.4: The acrylate based resins and formulations are mostly used for UV-coating applications with a market share of more than 80%. Polyester resins are mostly used in the wood coating industry and have a share of about 20% [5].

UV-curing process is mostly used. The following figures are the percentage of the market revenues in Europe in 2005 according to Frost and Sullivan, 2006 [5]:

- Printing and packaging: 32.8%

- Wood and Flooring: 28.7%

- Consumer goods: 19.9%

- Automotive: 10.1%

- Electronics: 8.5%

One third of the UV-coating industry revenues are turned over in the printing and packaging industry, which is highly connected to food packaging with strict regulations for migration levels of photoinitiators and product safety controls (see EU regulation No 1935/2004 6). This industry sector mostly uses offset or screen printing techniques, where as the latter is suited for being monitored by the presented sensor system.

Chapter 2

UV radiation curing of coatings and printing applications

The following chapter will introduce the emerging technique of ultra violet (UV) radiation cured polymer films. Coatings which are able to be polymerized using UV radiation have outstanding properties like surface hardness, chemical resistance and others compared to solvent based formulations. The polymerization process in the UV-curing is finished within milliseconds to seconds. The product is right afterwards ready for further processing. There is no volatile solvent in the formulation. Therefore parameters like viscosity or density must be controlled by additives or the amount of reactive monomer [7]. The latest developments in the automotive industry and printing industry for example have shown, that UV cured coatings and inks, especially acrylic based formulations, can successfully provide an alternative to conventionally solvent based formulations. Further details and informations can be found in Glöckner *et.al*, 2008 [8].

2.1 Photopolymerization

Photopolymerized coatings contain reactive chemical groups, which are linked together after energy-rich radiation (UV-radiation) has been applied onto the coating. This process is often started and maintained by radicals of the added photoinitiator (UV). In contrast to this, the electron beam curing (EB) is directly cross linking the coating without an added photoinitiator. Furthermore, the described curing processes (EB and UV) are realized without the help of any organic solvent and thus without the disadvantage of volatile organic compounds (VOC).

2.1.1   General composition of photocurable polymer formulations

A UV-curable coating or printing ink consists of mainly four components, oligomers, monomers, photoinitiators and some additives, like pigments or stabilizers. The chemical structures of the reactive components are displayed in Figure 2.2.

- **Monomer:** Typically monomers are for example Tripropylene Glycol Diacrylate (TPGDA) or 1,6-Hexanediol Diacrylate (HDDA). These monomers are used as the reactive part of the formulation and will be incorporated into the polymer network. Due to its low viscosity, they are as well used as a diluent in the coating. Percentage of monomer in the formulation: < 50.

- **Oligomer:** The compound is responsible for the actual backbone of the polymer network. Common oligomers are manifold modified epoxy acrylates or urethane acrylate resins, depending on the intended properties. Percentage of oligomer in the formulation: > 50 up to 80.

- **Photoinitiator:** The photoinitiator is responsible for the start and advance of the polymerization. It reacts with the monomers and oligomers and enables the chain grow of the polymer network. Two main types are mentioned here:

  - Radical photoinitiators: Free radicals are produced by UV-radiation. A type-I photoinitiator is mostly used, which undergoes a unimolecular process, called α-cleavage (Norrish Type 1). The α-cleavage process is shown in Figure 2.1.

  - Cationic photoinitiators: Protons or a Lewis acid is produced by UV-radiation.

  Percentage of photoinitiators in the formulation: < 3 usually less than 1.

- **Additives**: Components, which are responsible for the wetability or gloss. As well as pigments are added for a specific colour of the

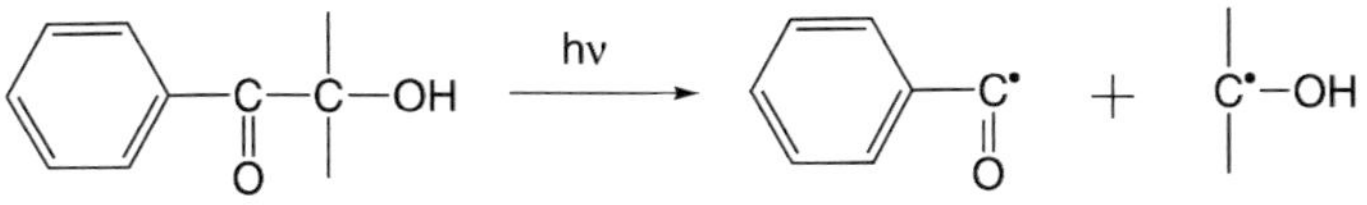

Figure 2.1: The α-cleavage (Norrish Type 1) process. The C$-$C bond adjacent to the functional group of the initiator is broken by the UV-radiation. The type-II photoinitiator is radicalised by a bimolecular reaction with a hydrogen abstraction, which involves a co-initiator. The efficiency of type-I is higher than that of type-II.

formulation. The additives do have a proportion of the formulation in the range of some weight percent. In the case of colour pigments the overall proportion is sometimes significantly higher and can reach up to a third of the final formulation.

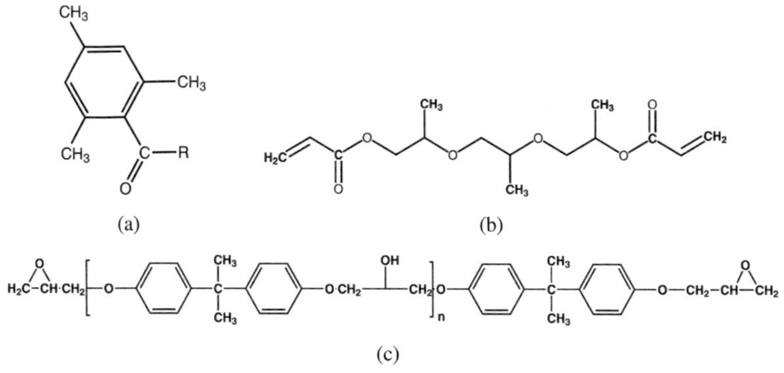

Figure 2.2: Chemical structure of main components in a UV curable polymer formulation. a) Radical photoinitiator (α-type I); b) Monomer (Tripropylene Glycol Diacrylate (TPGDA)); c) Oligomer (Modified epoxy acrylate)

2.1.2 Photopolymerization process

The photopolymerization is an addition reaction across the C$=$C double bond of the acrylate. Figure 2.3 shows the acrylate end group before and after the reaction. The cross-linking is realized by opening the C$=$C double bond, resulting in two open bonds. One bond is occupied by the photoinitiator radical and the second bond is used for the polymerization progress.

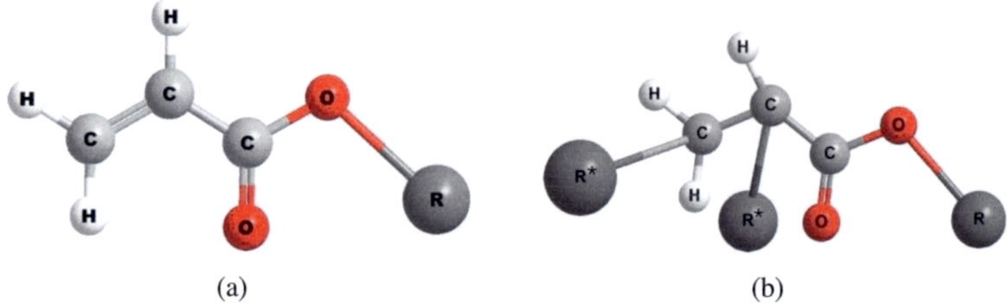

(a)   (b)

Figure 2.3: Acrylate end group of a photo-curable polymer before (a) and after (b) UV irradiation. The C-double bond in Subfigure (a) is not yet cracked. No linkage can occur, due to missing open bonds. Subfigure (b) shows the molecule after the photoinitiator has cracked the C-double bond. Two links are visible (R*). One link is occupied by the photoinitiator radical the other one is used for linkage to another molecule, forming the polymer network. The link (R) is connected to some other functional group within the monomer or oligomer.

The photopolymerization can be divided into several consecutive steps: Activation the photoinitiator by UV radiation, start of the photopolymerization, propagation of the polymer chain and final termination of the chemical reaction.

Figure 2.4 shows these main steps and explains the main changes in the chemical structures of Figure 2.2.

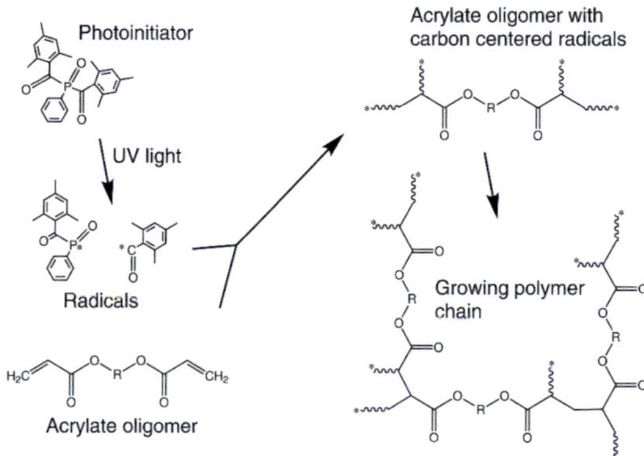

Figure 2.4: Example for the polymerization process. The photoinitiator is radicalised by a α-cleavage process [9]. These radicals will start the polymerization of the acrylate oligomer.

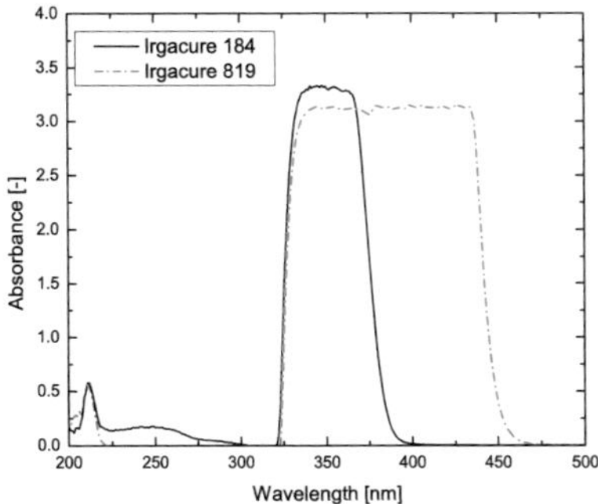

Figure 2.5: Absorbance spectra of industrial used type-I photoinitiators in the UV (5 wt% dissolved in acetone). A short wavelength initiator *1-Hydroxy-cyclohexyl-phenyl-ketone* (Irgacure184 ® by Ciba Chemicals, black line) and a second one *Bis(2,4,6-trimethylbenzoyl)-phenylphosphineoxide* (Irgacure819 ® by Ciba Chemicals, red, dotted line) [10].

1. Activating the photoinitiator [PI] by UV radiation, the Norrish Type-I initiator is cleaved into two radicals.

$$I \longrightarrow 2\,P \cdot \tag{2.1}$$

The initiation efficiency can be described as in Decker, 1998 *et al.* [11]

$$R_i = \Phi_i I_0 [1 - exp(-\epsilon l[PI])] \tag{2.2}$$

with
R_i = rate of initiation
Φ = quantum yield of initiation
I_0 = incident radiation intensity
ϵ = absorptivity
l = sample or coating thickness
$[PI]$ = concentration of photoinitiator

2. Start of the polymerization by the chemical reaction of the photoinitiator [9] radicals with the monomer and the oligomer. The C=C double bond at the acrylate end group is opened up and two links are produced. One link is saturated with the photoinitiator radical, see Figure 2.3. The second one is available for the chain growth

of the polymer. Therefore the polymerization reaction is not self-preserving, a steady supply of "fresh" radicals is necessary.

3. The polymer network is evolving. The activated oligomers and monomers are now adding up to the actual long-chained compounds. The propagation rate of this process can be influenced by controlling the temperature. Higher temperatures will lead to a slightly decreased polymerization time and a more developed polymer network [12], due to the lower viscosity and higher diffusion of radicals in the polymer network.

4. The chain growth can be finally terminated by several processes:

   - Recombination reaction of the radicals, as illustrated below, or two polymer chain ends couple together forming one long polymer chain.

$$P_m \cdot + P_n \cdot \longrightarrow P_{m+n} \qquad (2.3)$$

   - Disproportionation, by hydrogen abstraction, resulting in two non reactive polymer chains and one unsaturated end group.

$$P-CH_2-CH_2 \cdot + P-CH_2-CH_2 \cdot \qquad (2.4)$$
$$\longrightarrow P-CH2-CH3 + P-CH=CH_2$$

   - Trapping of radicals in the polymer network [13]. This effect causes the remain of uncracked monomers in the polymer and can lead to increased odour and less mechanical strength.

   - Oxygen inhibition: The produced free radicals are scavenged by the oxygen on the surface and in the upper layers of the coating [14, 15].

The final polymer network and coating thickness often exhibit shrinkage in a broad range of about 3 to 30% in comparison to the primary wet coating [16]. Notably this value can change easily within the same chemical formulation class, especially for industrially used coatings with high amounts of additives and pigments. The used formulation in this work (see Section 4.1) does show shrinkage of less than 10%.

2.2 Ultraviolet (UV) radiation sources for curing processes

2.2.1 Fundamentals of UV radiation

The definition of UV-radiation in the range of $100\ nm$ to $380\ nm$ is according to the German DIN 5031 part 7 [17] as follows:

- Vacuum UV-radiation (VUV): $100\ nm - 200\ nm$

- Short wave UV-radiation (UV-C): $200\ nm - 280\ nm$

- Medium wave UV-radiation (UV-b): $280\ nm - 315\ nm$

- Long wave UV-radiation (UV-C): $315\ nm - 380\ nm$

The VUV part is absorbed in the atmosphere by reacting oxygen to ozone $(O_2 + VUV = 2O \rightarrow O_2 + O = O_3)$. The UV radiation is commonly produced by plasma radiation sources. Mercury plasma discharge radiation is still the most effective way in producing UV-radiation. About 20% to 25% of the electrical power of a mercury medium pressure lamp is converted in to UV-radiation. The lamp filling consists of an inert gas and mercury, which is liquid at room temperature. The inert gas is ionised by the applied electric voltage and the power is heating up the mercury which vaporises. The vaporised mercury is further heated up and provides the radiating plasma. Due to the high proportion of mercury, the stabilised plasma produces a characteristic spectrum in the UV-range. The main resonant Hg-transition, hence most effective and significant emission line can be observed in the UV-C range at $253,7\ nm$ (see Figure 2.6(b)). By doping the pure mercury with iron, indium or gallium the emission spectra can be changed and tuned for a better fit with the absorbance spectra of the used photoinitiator (compare Figure 2.5 with Figure 2.6 (a) and (b)).

2.2.2 UV radiation sources

In most UV-curing applications Hg medium pressure lamps are used for the curing of various coatings, especially clear top coats. These lamps are suited to be installed on roll printing and coating machines as well as for three dimensional coatings of consumer goods or high quality industrial products. The lamps can vary in length from tens of centimetres up to two or more meters. In Figure 2.7 a typical Fe-doped lamp is shown, as used in small sized laboratory applications. The power consumption of these

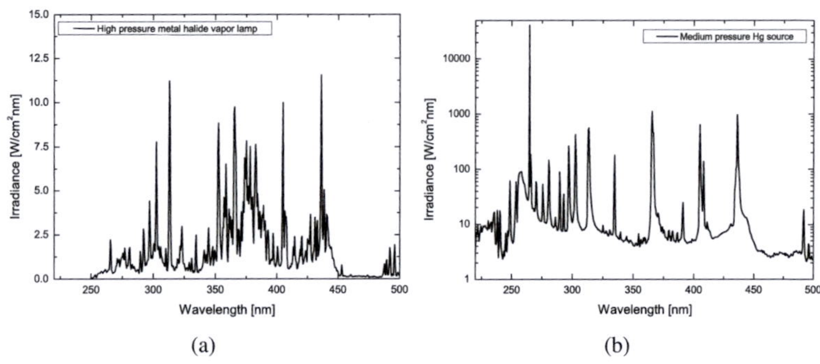

(a)   (b)

Figure 2.6: Emission spectra: (a) High pressure metal halide vapour lamp (Power: 400W); (b) Medium pressure mercury lamp (Power: 3kW) (b). See 2.2.2 for further details.

lamps is in the range of 1 kW to some tens of Kilowatts. In order to always ensure the required irradiation two or more lamps are arranged behind each other. Thus, the actual needed irradiance power for a proper curing is often more than doubled in the machine, in case of malfunction of one individual UV-lamp.

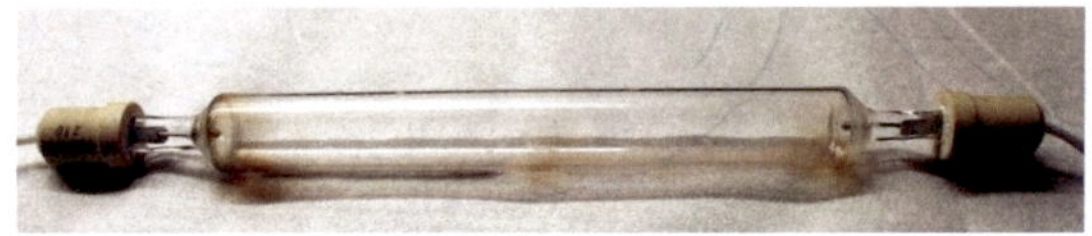

Figure 2.7: Fe-doped medium pressure mercury lamp (Power: 3kW), as used for industrial UV-curing applications like roll to roll screen printing. (Manufacturer: IST Metz GmbH)

The application of UV-curing in bonding processes, like electronics and optics (see Section 1.3), normally uses a high pressure metal halide vapour lamp for the curing process (see Figure 2.6). Due to its geometry, the radiation can easily be focused, so that an economical spot curing is possible. Table 2.1 compares the irradiance of the two discussed UV-radiation sources. The mercury medium pressure lamp irradiates about 68% of the radiation in the UV A, B and C range compared to the halogen-metal vapour lamp with only about 58% in the UV range. Notably the UV-C emission is reduced by nearly a factor of four in the metal halide vapour lamp, but the UV-A range is strongly increased.

As mentioned before, the benefits of the UV-C radiation can be only fully deployed with an almost oxygen free atmosphere [15]. Especially UV-C irradiation is responsible for the polymerization of the upper surface of a UV-coating. Therefore this type of lamp is preferred in industrial applications, where the surface hardness is most important, for example in wood panel coatings [18].

Latest developments of UV-Light emitting diodes (LED) are increasingly used in digital ink-jet printing applications, for example. The high flexibility of ink-jet printers and the efficient pinning and curing of the ink by UV-LED's are the main advantages compared to conventional printing systems [19]. However, the high price and rare wavelength availability put these advantages into question whether the UV-LED radiation sources will significantly replace the mercury plasma sources.

The production of high energetic UV-radiation (VUV) is realized by a Xenon-excimer source, mainly 172 nm. This kind of lamps are used for surface modifications for matting effects as often done in the wood coating industry [20]. Furthermore VUV radiation has a potential for water and air treatment. Flow-through photo-reactors have been examined, where the water is piped in a inner VUV transparent tube and the VUV plasma is produced in the outer tube [21]. Recently, Wang *et al.* [22] compared the disinfection effects of VUV radiation at 172 nm, 222 nm and 254 nm on spores in aqueous suspensions and could not prove a high efficiency using 172 nm radiation. Nevertheless, VUV radiation does have a higher efficiency in micro-organism deactivation compared to other disinfection systems based on UV radiation or ozone.

	Mercury medium pressure		Metal halide vapour	
	Irradiance $[W/m^2]$	Percentage	Irradiance $[W/m^2]$	Percentage
UV-A	2008.0	28%	154.5	38.9%
UV-B	1818.0	24%	59.5	14.8%
UV-C	1169.0	16%	19.0	4.7%
VIS	2346.0	32%	168.0	41.9%

Table 2.1: Emission distribution of the UV-sources displayed in Figure 2.6 according to the UV-range definition in DIN 5031, part 7 [17].

2.3 State of the art in quality control and process monitoring

The current quality control of UV-curing processes originates from conventional, solvent based printing applications. With this background custom quality tests for UV-cured coatings were developed. Still most quality tests only consider mechanical properties like scratch resistance and surface quality, e.g. acid resistance. Simple scratch and colour fastness tests are regularly performed by the printing machine operator right after production and before the product is finally packed. Most of the described methods in this section are standardized to ensure a proper comparison of the quality between different producers. The quality control tools mainly used by the printing and coating industry will be introduced below.

2.3.1 Mechanical quality test methods

2.3.1.1 *Methyl ethyl ketone (MEK) test*

The methyl ethyl ketone is rubbed across the coating surface multiple times using a cotton bud. The number of turns which are needed to uncover the substrate is used as a measure of the curing state of the coating. It verifies the proper curing of the top layers of the coating and ensures that the coating is solvent resistant. This test is specified in the German DIN 13523-11 [23] which regulates the MEK-Test for coil coated metals for example or more general in the U.S. standard specification ASTM D740-05 [24].

2.3.1.2 *Acetone wipe test*

An acetone soaked cloth is wiped across the coating surface. Unpolymerized parts of the surface will be dissolved and afterwards visible on the cloth. This method is very dependent on the wipe technique of the operator. To overcome this drawback acetone wipe test machines have been developed to ensure reproducible results. This method is used in the particular case of inner polymer coatings of aluminium tubes specified in the German DIN 15766 [25].

2.3.1.3 X-cut or right angle lattice cut and tape test

This test is a measure for the adhesion of the coating on a metal substrate, especially used for anti-corrosion protection or mechanical stressed areas of the coated part. It is made up of several perpendicular cuts of the coating with a sharp knife and removal of an afterwards applied pressure sensitive tape.

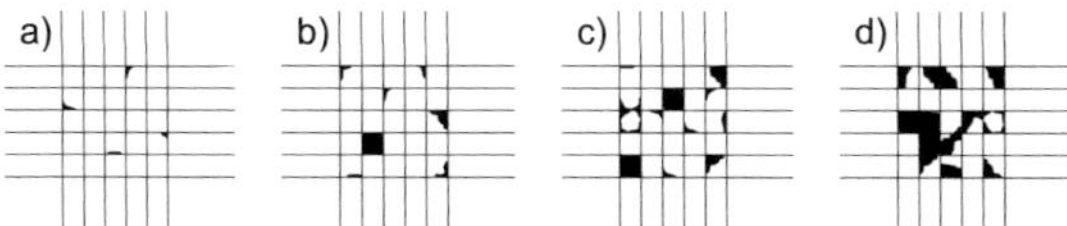

Figure 2.8: Right angle lattice cut test after removal of the tape. The black parts in the figures a) to d) show the damaged parts of the coating. The substrate is visible, which indicates a poor adhesion of the coating.

This test is specified in the US ASTM D3359-09e2 [26] regulation. The damage shown in Figure 2.8 of the right angle lattice caused by the removal of the adhesive tape is classified into 5 main levels described in Table 2.2. The reproducibility of this test can be increased by always deploying the same operator for all tests in one batch.

Classification	Pattern	Damage [%]
5	not displayed	0
4	a)	< 5
3	a) - b)	$5 - 15$
2	b) - c)	$15 - 35$
1	c) - d)	$35 - 65$
0	not displayed	> 65

Table 2.2: Classification of adhesion test results of the right angle lattice cut and tape test according to US ASTM D3359-09e2 [26].

The right angle lattice cut and tape test is most popular in the automotive industry with its component suppliers. They often apply multi layered coating systems in which the adhesion is often a critical property.

2.3.1.4 Pendulum damping test after Koenig or Persoz

The pendulum damping test after Koenig or Persoz mechanically measures the surface hardness of the coating. A tungsten carbide sphere on a pendulum arm is deflected by several degrees and swings back on to the coating surface. The number of swings in between two well defined angles is counted. The angle limits are defined for the Koenig hardness in the range of $3°$ to $6°$ and $4°$ to $12°$ for the Persoz hardness. With increasing hardness of the surface, the swings in between the angle limits are increasing as well. The described two methods hardly differ; some minor differences are in size and weight of the spheres and the individual mechanical set-up. These methods are standardized in the specification ISO1522 [27].

2.3.2 Analytical quality test methods

The above describe test methods (Subsection 2.3.1) are only indirectly examining the degree of conversion in the UV-cured coating. Physical properties like surface hardness or solvent resistance are used to decide if a sufficient quality is achieved. Analytical test methods like FT-IR spectroscopy or photospectroscopy, which are briefly introduced in this section, can be used for gaining direct information about the chemical properties of the coating. Due to the molecular changes in the polymer formulation as described in Section 2.1.2 , different spectra can be measured.

2.3.2.1 FT-IR Transmission and Attenuated total reflection (ATR) spectroscopy

Spectroscopy in general is a powerful tool for detecting specific chemical compounds by identifying the characteristic molecule vibrations mainly in the Mid-Infrared range (MIR). Most laboratory based spectroscopes use Fourier-Transformed signals, due to the low signal amplitudes in the MIR. For example, a Michelson-Interferometer is used for scanning the absorption in the desired wavelength range and a Fourier transformation is performed from the time domain to the frequency domain. The Attenuated-Total-Reflection principle (ATR) is used in spectroscopy as a signal enhancer and enables the user to examine liquid, solid or gaseous samples very accurate with little effort. The radiation is guided in a material with a high refractive index (>2.4 to 4), for example diamond. The attenuated total reflection at the interface of the sample and the diamond, produces an evanescent field which penetrates into the sample by 0.5 to $3.0\,\mu m$. The radiation is guided back into the spectrometer and onto the detector. This technique is often used for scientific analysis of all kinds of materials, due to its high precision and simplicity. A detailed description of the above mentioned principles can be found in Chapter 3 (see also Figure 3.3).

2.3.2.2 *UV-VIS-NIR Photospectroscopy*

Photospectroscopy in the near infrared (NIR) or UV-region is used for the quality control of single constituent like photoinitiator absorbance or colour pigment appearance. In contrast to the mid and far infrared most UV-coatings do only have few or even none characteristic absorption patterns in the NIR. Photospectroscopy on UV-curing coatings is often limited to fundamental research at universities and scientific facilities. Details on photospectroscopy can be found in Chapter 3 Section 3.4.2.

Chapter 3

IR-Spectroscopic methods for determination of conversion

Infrared spectroscopy is a powerful tool for characterizing molecules and atoms in terms of structure and constituents. In the year 1800 Sir William Herschel discovered the existence of infrared radiation next to the visible spectrum. Since then the infrared spectrum gained importance in many areas of today's science.

Polymer science is strongly based on the analysis of new compounds and their chemical structure with the infrared spectroscopy technique. The range between $2000\ cm^{-1}$ and $500\ cm^{-1}$ is often called the fingerprint area of organic compounds. This region reveals unique absorption patterns of many polymers and organic solvents.

Spectroscopic measurements in this finger print spectrum are simple to be performed with today's laboratory equipment and provide reliable analysis results on purity or single constituents within short time.

Molecules and atoms can interact with electromagnetic radiation by absorbing the energy, resulting in a transition to an excited state of the molecule or atom with electrons in higher orbits. The absorption of electromagnetic radiation is only possible at distinct energy levels, or frequencies, dependent on the molecular structure and bounded elements in the molecules. This behaviour can be differentiated in vibrational modes, for example stretching and rocking or rotational modes. The transition energies of rota-

tional modes are some magnitudes lower than the energies of the vibrational modes and thus they are not as predominant in the Infrared spectrum.
A detailed description of the vibrations can be found in Section 3.1 of this Chapter. As an excited state of the molecule is normally unstable the same or smaller amount of energy, which has been absorbed, will be released by the molecule as electromagnetic radiation again.

3.1   Characteristic vibrations of molecules [1]

A FT-IR spectrum reveals frequencies, at which absorption in the examined molecule occurs. From the beginning there is no information on any specific chemical compound, chemical group or even on a specific atom oscillation. The observer is only measuring an energy transition at a distinct frequency. To connect chemical compounds or reaction behaviour of the unknown material to the FT-IR spectroscopy results, a deeper understanding of the physical processes, taking place in the molecule is necessary. Several descriptive models for the vibrations of molecules have been developed as well as analytical and numerical methods for describing the physical properties of the inner molecular processes. The following chapter will introduce some of these models.

3.2   Fundamental vibrational modes in polymers and its approximation

The basic understanding of the mechanisms connected to molecular absorption can be described by the possible movements of a molecule or part of a molecule and its atoms. A simple part of a molecule is the $-CH_2-$ group (Methylene) for example, which is present in many polymers or other organic compounds. This simple chemical compound is used as a descriptive example of the possible molecular vibrations and rotations.

The fundamental vibrational modes of molecules or their chain parts can be described by the following atom movements. The described vibrations are pictured in Figure 3.1. A brief derivation of these modes is done in section 3.2.1.

- Symmetrical stretching:
  The bond length between the C molecule and H molecules is sym-

---

[1]Rotational modes which are far less dominant in the infrared region are not in the scope of this work.

metrically changed.

- Asymmetrical stretching:
  The bond length between the C molecule and H molecules is asymmetrically changed.

- Scissoring:
  The angle between the C molecule and H molecules is changed. Both H molecules change the same angle in opposite direction.

- Rocking:
  The angle between the C molecule and H molecules is changed. Both H molecules change the same angle in the same direction.

- Wagging:
  Symmetrical out of plane vibration of the H molecules.

- Twisting:
  Asymmetrical out of plane vibration of the H molecules.

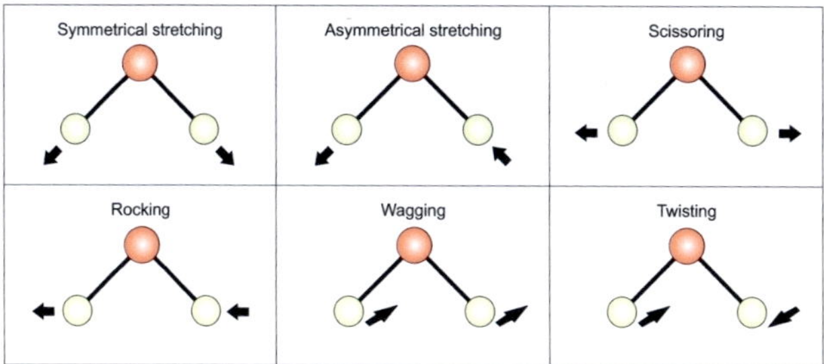

Figure 3.1: Idealized fundamental vibrations of molecules or molecule parts. All these vibrations may be superimposed by each other.

3.2.1 Dynamics of small molecules: Approximation of the vibrational potential

The mathematically exact and physical elaborate way describing molecule movements is far beyond the scope of this work. Hence only the main equations and assumptions, used to describe molecular vibrations, will be summarized in the following section. Continuative information on this topic can be found in Wilson *et al.*, 1955 [28].

To describe a molecule and its vibration, it has been found useful to apply the principle of small vibration. In Cartesian coordinates x_n, y_n, z_n with respect to the centre of mass of the n-th atom and with the equilibrium coordinates a_n, b_n, c_n the kinetic energy T can be written as in Wilson *et al.*, 1955 [28]:

$$2T = \sum_{n=1}^{N} m_n \left[\left(\frac{d\Delta x_n}{dt} \right)^2 + \left(\frac{d\Delta y_n}{dt} \right)^2 + \left(\frac{d\Delta z_n}{dt} \right)^2 \right] \tag{3.1}$$

with $\Delta i_n = i_n - j_n$; $i = x, y, z$; and $j = a, b, c$ as the displacement from the equilibrium, respectively and m_n as the mass of the n-th atom. For convenience the coordinates are transformed to mass weighted coordinates $q_1, q_2, ..., q_{3N}$ as $q_1 = \sqrt{m_1}\Delta x_1$; $q_2 = \sqrt{m_1}\Delta y_1$; $q_3 = \sqrt{m_1}\Delta z_1$; $q_4 = \sqrt{m_2}\Delta x_2$; $\cdots$.

If we use these coordinates in Equation 3.1 the kinetic energy T states as the derivative of q_i :

$$2T = \sum_{i=1}^{3N} \dot{q_i}^2 \tag{3.2}$$

If we assume small displacements, the potential energy V is a function of q and can be written as a Taylor expansion truncated after the second order:

$$2V \approx 2V_0 + 2 \sum_{i=1}^{3N} \left(\frac{\partial V}{\partial q_i} \right)_0 q_i + \sum_{i,j=1}^{3N} \left(\frac{\partial^2 V}{\partial q_i \partial q_j} \right)_0 q_i q_j$$

$$2V \approx 2V_0 + 2 \sum_{i=1}^{3N} f_i q_i + \sum_{i,j=1}^{3N} f_{ij} q_i q_j \tag{3.3}$$

Where the index 0 indicates the equilibrium state of the system.

With the assumption, that the energy is zero at the equilibrium state of the molecule we can eliminate V_0. If in addition all displacements q are zero, the energy must have a minimum:

$$\left(\frac{\partial V}{\partial q_i} \right)_0 = f_i = 0 \ \text{ with } \ i = 1, 2, 3, ..., 3N \tag{3.4}$$

Considering equation 3.4, a solution to equation 3.3 can only be given analytically. Thus it states:

$$2V = \sum_{i,j=1}^{3N} f_{ij} q_i q_j \ \text{ with } \ f_{ij} = f_{ji} = \left(\frac{\partial^2 V}{\partial q_i \partial q_j} \right)_0 \tag{3.5}$$

To finally solve the above stated equation Newton's law of motion is used:

$$\frac{d}{dt}\frac{\partial T}{\partial \dot{q}_j} + \frac{\partial V}{\partial q_j} = 0 \ \text{ with } \ j = 1,2,3,...,3N \tag{3.6}$$

As defined before, T is only depended on the velocity. Thus equation 3.5 describes an harmonic oscillator:

$$\ddot{q}_j + \sum_{i=1}^{3N} f_{ij} q_i = 0 \ \text{ with } \quad j = 1,2,3,...,3N \tag{3.7}$$

As equation 3.7 is known to be a second order linear differential equation, an analytical solution can be found:

$$q_i = A_i \, cos(\omega t + \phi) \tag{3.8}$$

with the amplitude A_i, the frequency ω and the arbitrary phase ϕ. By substituting equation 3.8 in equation 3.7, the set of $3N$ equations

$$\sum_{i=1}^{3N} (f_{ij} - \delta_{ij}\omega^2) A_i = 0 \ \text{ with } \ j = 1,2,3,...,3N \tag{3.9}$$

is used to determine the amplitudes A_i. Here, δ is the Kronecker symbol. Only for certain values of $\omega_k = \omega$ the solution is non-trivial and the determinant of the equations 3.9 vanishes:

$$\begin{vmatrix} (f_{1,1} - \omega^2) & f_{1,2} & f_{1,3} & \cdots & f_{1,3N} \\ f_{2,1} & (f_{2,2} - \omega^2) & f_{2,3} & \cdots & f_{2,3N} \\ \cdots & \cdots & \cdots & \cdots & \cdots \\ f_{3N,1} & f_{3N,2} & f_{3,N3} & \cdots & (f_{3N,3N} - \omega^2) \end{vmatrix} = 0 \tag{3.10}$$

For a fixed value ω_k, the set of equations is solved for the corresponding A_i^k. Notably, only the ratios of the amplitudes A_i^k / A_j^k can be determined within this system of equations.

The A_i^k are scaled to $\tilde{A}_i^k$, in order to determine the normalized quantities l_{ik}:

$$l_{ik} = \frac{\tilde{A}_i^k}{\left[\sum_i (\tilde{A}_i^k)^2\right]^{\frac{1}{2}}} \ \text{ with } \ \sum_i l_{ik}^2 = 1 \tag{3.11}$$

A physically correct solution may be obtained by introducing the constant K_k to determine finally the normalized coordinates q_i as

$$q_i = A_i^k \cos(\omega_k t + \phi) = K_k l_{ik} \cos(\omega_k t + \phi) \qquad (3.12)$$

The determinant 3.10 characterizes the vibrational behaviour of molecules. The following description in Section 3.3 is based on this relation.

3.3 Fundamental modes and overtone vibrations

3.3.0.1 *Fundamental modes*

The harmonic oscillator solution in equation 3.8 for the molecule movement with frequency $\omega_k/2\pi$ and phase ϕ_k fulfils the requirements for an oscillation of each atom around its equilibrium. Only oscillations with equal phase and frequency of each coordinate $q_1, q_2, ..., q_{3N}$ are known as a normal or fundamental mode of oscillation (see Figure 3.1). A detectable IR absorption only occurs, if the motion includes a change in the electric dipole moment. The frequency of a classical harmonic oscillator can be calculated to:

$$\omega_k = \frac{1}{2\pi}\sqrt{\frac{K}{\overline{m}}} \qquad (3.13)$$

with $\overline{m}$ for the reduced masses of the bond forming atoms and K for the force constant.

3.3.1 Overtone modes

A thorough understanding of overtone modes in molecule bonds needs a deep understanding of quantum mechanics as described in Wilson *et al.*, 1955 [28] or more recently for computational use in Zerbi *et al.*, 2007 [29]. The overtone modes are always a multiple of the underlying fundamental mode. Quantum mechanics provide a way to calculate for those overtones v_ν:

$$v_\nu = [1 - (\nu + 1)\,x]\nu\,\omega_0 \qquad (3.14)$$

where ν is the vibrational quantum number, restricted to only positive integers $\nu = 1, 2, 3, ...,$ x the anharmonicity constant and ω_0 the fundamental frequency. As speaking of first, second and third overtones the quantum number ν is 2, 3 and 4, respectively. Equation 3.14 is only valid for

simple molecules for example chloroform ($CHCl_3$), with little complexity. Complex polymer chain molecules can not be approximated by the above equation. A summary on recent results in calculating overtone spectra of complex molecules is given by Pavlyuchko *et al.*, 2011 [30].
Additional work on fundamental and overtone modes of polymers in the Infrared can be found in Wheeler, 1959 [31] and even more profound in Fox, 1940 [32].

3.4 Absorption spectroscopy

The spectroscopic analysis of chemical compounds has been performed since the 1930s. Kirchhoff and Bunsen were the first to pioneer the absorption spectroscopy of flames and to explain the origin of the Frauenhofer lines (Details in Koirtyohann, 1980 [33]).
Today's application of absorption spectroscopy almost affects any research area dealing with organic materials. This fact also influences the development of cost-effective spectroscopic equipment, which leads to an increased understanding of the basic principles even in lower production levels.

3.4.1 Fourier Transform Infrared (FT-IR) spectroscopy

Standard FT-IR spectrometers are using a Michelson interferometer to overcome the lack of low radiation powers of far infrared sources, as the often used globars. The radiation of the source is modulated by a moving mirror and superimposed by a beam splitter in the Interferometer. After passing the sample, the detected signal (interferogram), which composed of all desired frequencies, is processed by a Fast Fourier Transformation with respect to the modulating frequency. The detected signal is transformed from the space domain to the time domain. The main motivation of this technique is to get a maximum amplitude on the detector for a increase of the signal to noise ratio.
In contrast to this approach, common absorption spectrometers select the desired frequency before passing the sample, with a notable lower signal amplitude and thus higher noise level in the result.
An overview on the used materials for the beam splitter and possible detectors for the far and mid infrared is given in Figure 3.2 (a), Figure 3.2 (b) shows the principle set-up of a often used Michelson Interferometer.

A interferogram can be described in its simplest way as

$$I(\delta) = A(\nu)cos(2\pi\nu\delta) \tag{3.15}$$

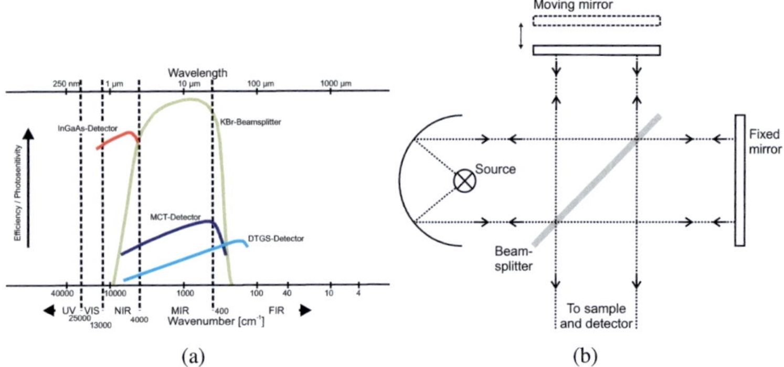

Figure 3.2: (a) Overview on beam splitters, detectors used in the mid and far infrared in FT-IR spectrometers (source: Bruker Optics). (b) Michelson Interferometer, one mirror is used for modulating the incoming light. The motion causes a change in the optical path length and thus to an interference of the reflected light at the beam splitter. This method provides the frequencies in the space domain with a high signal amplitude and is used for increasing the signal to noise ratio by some orders of magnitude.

with $I(\delta)$ as the cosine Fourier transform of the interferogram, $A(\nu)$ the beam intensity at a wavenumber ν and δ as the beam retardation of the Michelson interferometer. Yet, most interferometer do a modulation of the mirror with a constant velocity v, therefore the retardation δ can be written in terms of time and velocity as

$$\delta = 2vt \quad [cm] \tag{3.16}$$

and the cosine Fourier transform from equation 3.15 becomes

$$I(t) = B(\nu)cos(4\pi\nu \cdot vt) \tag{3.17}$$

Equation 3.17 now represents the Fourier Transform from the space or wavenumber domain to the time domain. The calculated Fourier Transformation does not normalize the amplitudes of the spectra, the amplitude results are further processed for proper results.

A standard method is the vector normalization of each of the spectra to get comparable results. Special care has to be taken for the quality of the acquired interferogram and the modulating frequency depending on the application. Furthermore noise and unwanted atmospheric absorption windows, like CO_2 or H_2O have to be taken into account for processing.

More details on the classic FT-IR theory and its application in Michelson Interferometer can be found in Griffiths *et al.*, 1986 [34].

3.4.1.1 *ATR-FTIR spectroscopy*

The basic of the attenuated total reflection spectroscopy is the fact, that light propagating in a optically dense medium with a refractive index n_c undergoes a total internal reflection at an interface to a medium with a lower refractive index n_s ($n_s < n_c$) in case the incident angle of the light is above the critical angle for the total internal reflection. The resulting electromagnetic phenomena of this effect is called an evanescent field. As mentioned, it can only exist above a critical angle $\theta = sin^{-1}\frac{n_s}{n_c}$ [35]. The basic principle and the ATR set-up is displayed in Figure 3.3.

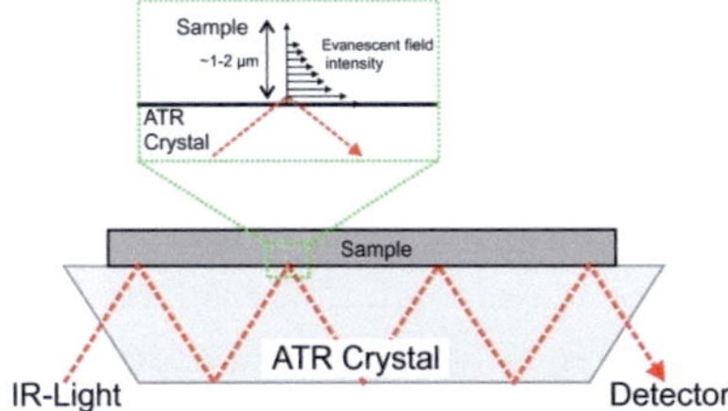

Figure 3.3: ATR-principle in absorption spectroscopy. The IR-light is guided in the ATR-crystal. A part of the resulting evanescent field is absorbed in the sample, depending on the sample properties.

The penetration depth of the evanescent wave into a sample is obtained from the electric field in the medium with $n_s << n_c$, which decays as:

$$E = E_0 \, exp \left[\frac{-2\pi}{\lambda} \left(sin^2\theta - \left(\frac{n_s}{n_c} \right)^2 \right)^{\frac{1}{2}} z \right] \qquad (3.18)$$

with E_0 as the initial electric field amplitude; $\lambda = \lambda_{vac}/n_c$ represents the wavelength of IR-light in the denser medium; θ is the angle of incidence; z defines the distance from the interface. Equation 3.18 can be written as

$$E = E_0 \, exp\left(-\gamma z\right) \qquad \text{with} \quad \gamma = \frac{2\pi\left(sin^2\theta - \frac{n_s}{n_c}^2\right)^{(1/2)}}{\lambda} \qquad (3.19)$$

Harrick and du Pre, 1966 [36] defined the parameter "depth of penetration" d_p as the depth, where the amplitude of the electromagnetic field decays to $E_0 \, exp^{-1}$. Thus with $z = d_p = \frac{1}{\gamma}$, we can solve for d_p:

$$d_p = \frac{\lambda}{2\pi}(n_c^2 sin^2\theta - n_s^2)^{\frac{1}{2}} \qquad (3.20)$$

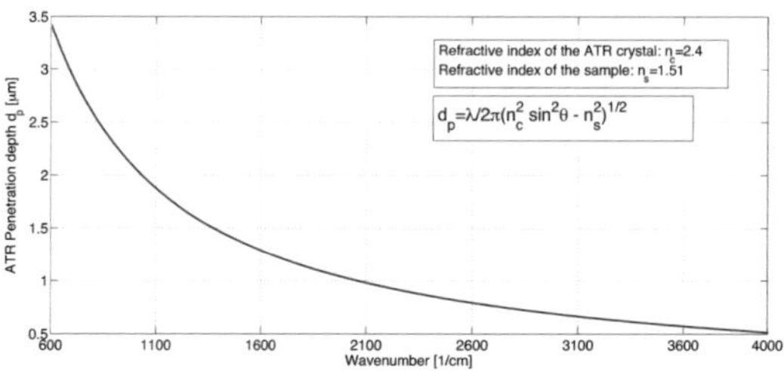

Figure 3.4: Penetration depth of the evanescent field into a sample with a refractive index of $n_s = 1.51$ and an ATR crystal with $n_c = 2.4$ as a function of the wavenumber.

Figure 3.4 shows the penetration depth as a function of the wavenumber for a refractive sample index $n_s = 1.51$. In the range between $600\,cm^{-1}$ and $2000\,cm^{-1}$ the penetration depth is calculated to be between $1.0\,\mu m$ to $3.0\,\mu m$. The discussed wavelength range is most important for polymer spectroscopy, due to the fundamental modes observed in this region. Further information on the ATR principle can be found in Harrick *et al.*, 1965 and 1966 [37, 36], where the first fundamental work on the topic of internal reflection spectroscopy in bulk materials is presented. A conclusive review on the principle of the internal reflection spectroscopy (ATR spectroscopy) can be found in Mirabella *et al.*, 1985 [35].

3.4.2   UV / VIS / NIR absorption and reflection spectroscopy

Photospectroscopy is a widely used method for the characterization of solid and liquid materials in the wavelength range between $180\,nm$ in the deep UV to about $3300\,nm$ in the Mid-IR. For example pharmaceutical companies, as well as optic producers are using photospectroscopy to acquire concentrations of agents in solutions or transmission coefficients of their products.

The high accuracy and fast data acquisition times of state of the art spectrophotometers are the main reasons why spectrophotometers are preferred against time consuming and expensive chemical analysis. Figure 3.5 shows a transmission measurement on different $CuSO_4$ solutions, used for the determination of the $CuSO_4$ concentration in H_2O.

 Furthermore the development of organic solar cells and organic LED's are strongly relying on the fundamental results on transmission and absorption measurements on new materials in the visible spectrum.

The basic principle of a spectrophotometer is the comparison of two

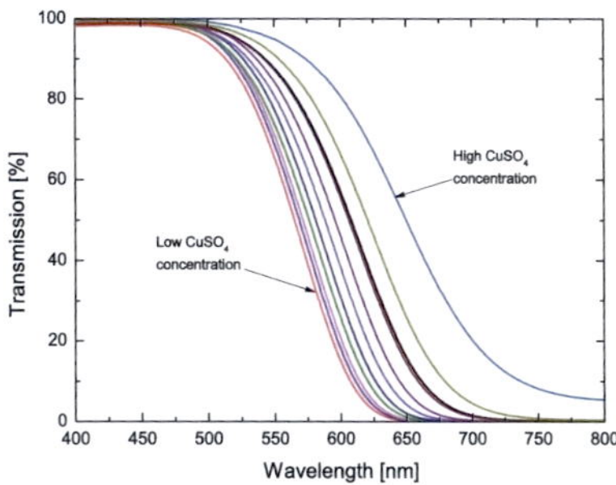

Figure 3.5: Transmission of solutions with different concentrations of $CuSO_4$. A typical wavelength range for photospectroscopy from $400\,nm$ to $800\,nm$ is used.

monochromatic light beams, a reference and a sample beam; where as the light beams are generated from one source by means of a reflective light chopper. This resulting transmittance is the quotient of two intensities, the intensity of the reference beam and the intensity of the transmitted sample beam.

$$T = \frac{I_{Sample}}{I_{Reference}} \tag{3.21}$$

This method has been enhanced in recent years, and is now capable to measure optical densities of up to $OD\,7$. In this work the state of the art spectrophotometer "LAMBDA 1050" from Perkin Elmer is used in combination with an integration sphere, which enables the user to measure reflection of surfaces and scattering samples.

3.4.2.1 Spectrophotometer Lambda1050 from Perkin Elmer

The spectrophotometer Lambda1050 from Perkin Elmer is capable of characterizing samples in terms of their optical transmission and reflection between the deep UV ($175\,nm$) and the NIR/MIR ($3300\,nm$). It is equipped, if used in the "'transmission only"' mode, with three detectors, a photomultiplier ($180\,nm$ - $\approx$ $850\,nm$), an InGaAs-detector ($\approx$ $850\,nm$ - $\approx$ $1900\,nm$) and a PbS-detector for the range up to $3300\,nm$. As light sources, deuterium (UV) and tungsten (VIS-NIR) lamps are used. The modular set-up of the spectrophotometer enables the user to quickly change between the "'transmission only"', the "'integration sphere"' and the "'absolute direct reflection"' modules. The maximum optical resolution of the

Lambda1050 is defined by the manufacturer to $\leq 0.05\,nm$ and $\leq 0.20\,nm$ in the UV-VIS and NIR range, respectively [38].

3.4.2.2 Integration sphere of the Lambda 1050

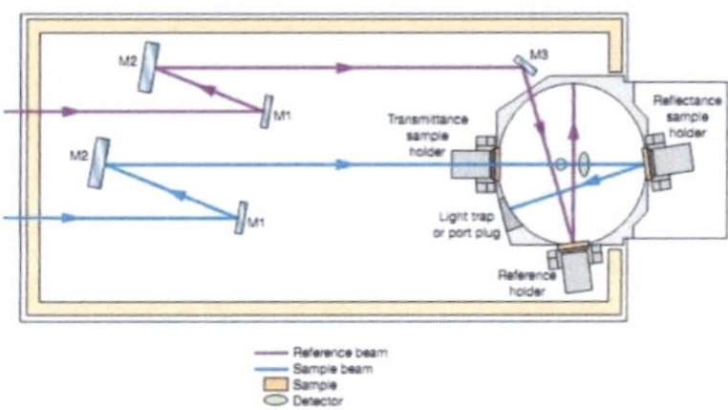

Figure 3.6: Design of the Perkin Elmer Integration sphere used for the reflection reference measurements of all coating samples. The used integration sphere is equipped with an photomultiplier and an InGaAs-detector. The transmission entrance port, the reflection port, the light trap and the reference port are shown as well. (Adapted from Perkin Elmer [39])

The integration sphere used in the laboratory work is made of spectralon. Spectralon exhibit the best known diffuse spectral reflection over a range from about $250\,nm$ up to $3300\,nm$. Therefore it is best suited to be used in this kind of applications. The integrating sphere is designed for transmission and reflection measurements with two different set-ups: Collecting the diffuse and direct light or collecting only the diffuse reflected light within the sphere. The latter set-up is implemented by a light trap at that point, where the deflected direct light hits the sphere (see Figure 3.6).
Within this set-up absolute measurements of the reflection and transmission properties of scattering samples are possible.

3.5 Emission spectroscopy

Characterizing radiation sources in terms of spectral power distribution is most important for the development of new light sources and their possible benefit for efficient energy usage. Emission spectroscopy is routinely used for the determination of the effectiveness of gas discharge lamps. Different gas constituents can be analysed in the excited state. With the knowledge of the physical behaviour of plasmas and these measurements, critical parameters like gas pressure can be calculated and optimized. Furthermore in recent years the power efficiency and spectral light distribution of organic

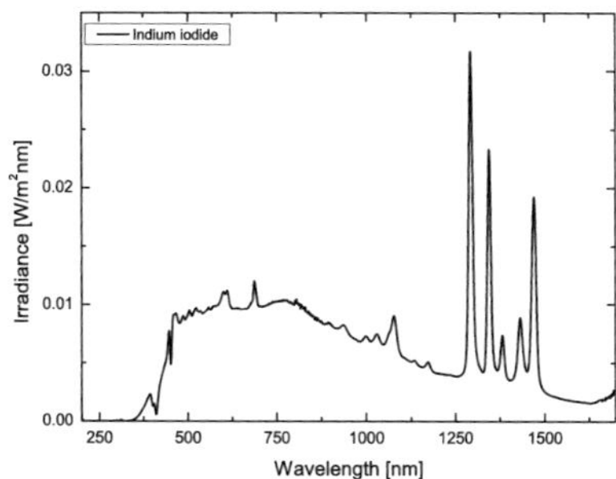

Figure 3.7: A typical example for emission spectroscopy: Indium Iodine high pressure plasma exited by microwave radiation. (Data courtesy of Christoph Kaiser, LTI, KIT)

light emitting diodes is getting into the focus, for example Gather *et al.*, 2011 [40].

Another new approach within the gas discharge lamps is the approach to excite the plasma with the use of microwave radiation. This approach is promising for the development of mercury free discharge lamps with similar performance values as today's mercury containing light sources.
In Figure 3.7 an emission spectra of a high pressure lamp is shown, excited by microwave radiation. Industrial combustion applications are monitored by emission spectroscopy for process optimization and controlling. This application determines the concentration of critical exhaust gases or temperatures directly in the combustion chamber and is a most important tool for the prevention of far reaching environmental consequences, for example in refineries or power plants [41].

Chapter 4

UV-cured coatings

This chapter includes the preparation and spectral characteristics of the custom model formulations as well as the coatings being already used in industrial processes. In regards to prove the basic functionality of the sensor system, a simple coating with the essential components of a UV-coating has been developed. After successful tests with the model formulation, the next step is taken by analysing more complex coating formulations supplied by the industry. These coatings are already containing all components necessary for the full functionality in terms of adhesion, gloss and surface protection matters and other desired properties.

The formulations and coatings are characterized by using the IR techniques explained in Chapter 3. FT-IR ATR and NIR spectra are used to measure the polymerization process and the curing state. The different absorption peaks of the acrylate components are discussed and traced down from the far infrared to the near infrared in accordance with the oscillation and absorption principle discussed in Sections 3.2 and 3.3.

4.1 TPGDA-epoxy acrylate model formulation

Various kinds of formulations were prepared during the development of the optical sensor system. The following section will only introduce those formulations, which were finally used for the scientific work. The formulation will be characterized in terms of characteristic IR absorptions and the associated overtones in the Near-Infrared.

The clear coating for most sensor measurements and photometric verification is composed of modified epoxy acrylate resin (LAROMER
LR9019 [42], $75\,wt\%$) and tri(propyleneglycol)diacrylate (LAROMER
TPGDA [43], $25\,wt\%$) purchased from BASF Laromer. The viscosity of
LAROMER LR 9019 and LAROMER TPGDA is specified to $15-25\,Pas$
at $23°C$ and $10\,mPas$ at $25°C$, respectively. The ratio of $75:25$ of LR
9019 and TPGDA was found to be the best ratio for the application with a
spiral doctor blade.

The white pigmented coating is made by dispersing $2\,wt\%$, $5\,wt\%$ or
$10\,wt\%$ of titanium dioxide (TiO_2, KRONOS 2059) in the clear formulation, respectively. The average size of the TiO_2 pigments has been determined to $2\,\mu m$ diameter by means of scanning electron microscopy (SEM)
(see Figure 4.2).

The photoinitiator Phenyl-bis(2,4,6-trimethyl benzoyl) phosphine oxide
IRGACURE 819 [10] from Ciba Speciality Chemicals is used with a concentration of $3\,wt\%$ per formulation (see Figure 2.5 for more details). The
high ratio of the photoinitiator ensures maximum polymerization of the formulation. A resulting drawback is a slightly yellow colour of the transparent formulation, which will be normally avoided in industrial applications.

4.1.1 Preparation of the formulations

All formulations are prepared by the right mass of each component. A scale
with a resolution of $\pm0,001\,mg$ is used. All components are agitated with
a magnetic stirrer for at least 24 hours to ensure a sufficient mixing (see
Figure 4.1). To protect the photosensitive formulation against unwanted
photopolymerization, all work is performed in a UV protected environment.
Due to the limited laboratory capacities, $50\,g$ of the final formulation are
prepared at once, which equals $\approx 45\,cm^3$ at a density of $1.1\frac{g}{cm^3}$. The
average coating volume needed for one steel panel sample (maximum
thickness $\approx 80\,\mu m$; size $50\,mm$ x $90\,mm$), is calculated to $0.36\,cm^3$. With
an average sample number of 10 - 12 per measurement batch, this amount
is sufficient to ensure equal conditions and consistent data quality in each
run. The shelf life of the formulation is calculated analogue to the shelf life
of the commercial components, which is defined to 6 months.

The TiO_2 pigments were carefully added into the final clear formulation
and stirred again for more than 48 hours. Less mixing time showed an in-
sufficient distribution of the TiO_2 particles, observed during the application

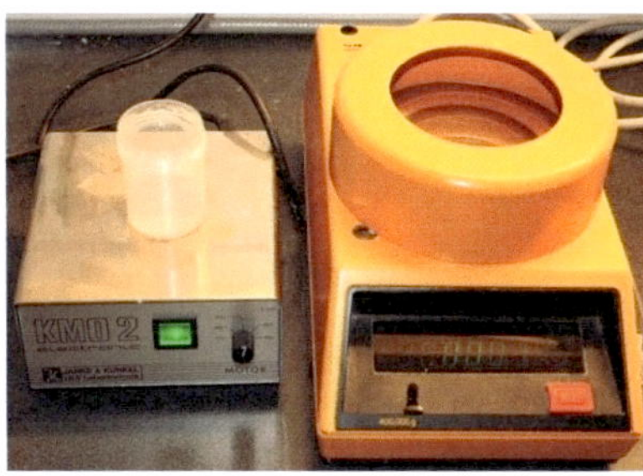

Figure 4.1: Laboratory equipment used for the preparation of the model formulation. Left: Magnetic stirrer; Right: precision scale;

of the formulation. A proper pigment distribution in a polymerized coating is shown in Figure 4.2.

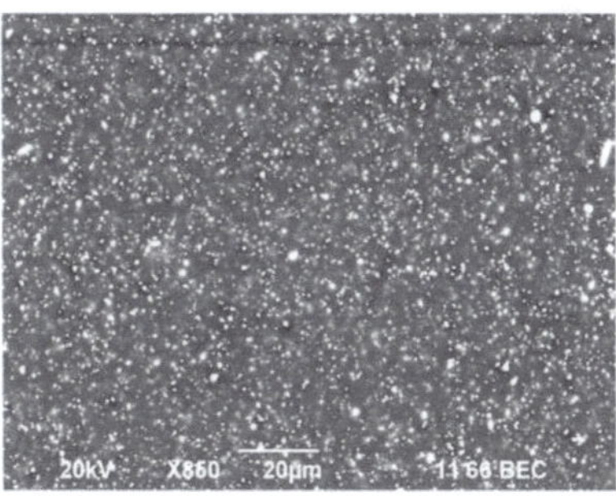

Figure 4.2: TiO_2 pigments (concentration of 5 wt%) in the clear polymerized formulation. The pigment size is calculated to an average size of $2\,\mu m$. The SEM picture clearly shows a proper pigment distribution within the coating layer.

The measurements in Section 6.2 are performed with TPGDA mixed with *Aerosil 200* to increase the viscosity. *Aerosil 200* is the trade name of a hydrophilic pyrogenic silicic acid (Dioxosilane O_2Si). The specific surface of *Aerosil 200* is $200\,m^2/g \pm 25$ with an average particle size of $12\,nm$. *Aerosil 200* is often used by the coating industry to adjust the rheology and thixotropy of various coatings, inks and sealants.

4.1.2   Characteristic oscillations and overtones in the IR

The main molecular oscillations of acrylic photocurable polymer formulations are described in the following paragraph. The fundamental modes of the chemical structure of the acrylate before and after photopolymerization are identified and labelled according to Sections 3.2, 3.3 and fundamental publications.

The model formulation, described in 4.1.1 shows acrylic oscillations between $800\,cm^{-1}$ and $1720\,cm^{-1}$. At $1721.1\,cm^{-1}$ the C=O acrylic carbonyl stretch vibration is clearly visible in Figure 4.3 labelled with index 6. This chemical binding does not change during the photopolymerization process and is therefore used as a normalization value to ensure proper comparison of related acrylate coating samples, especially unpolymerized and polymerized ones.

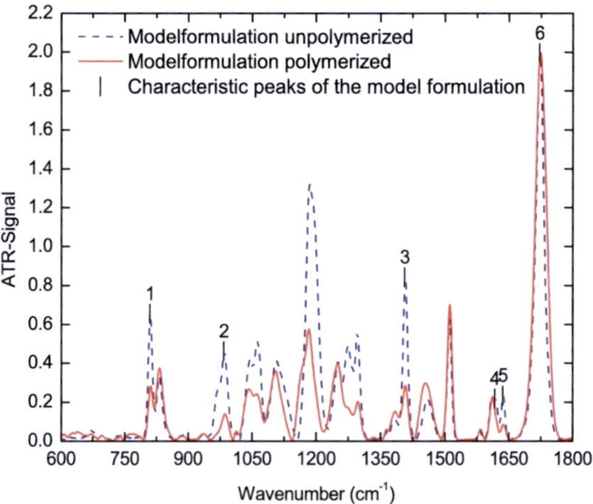

Figure 4.3: ATR-FTIR spectra of the model formulation used in the experiments, before and after UV curing. The labelled absorbance peaks are described in Table 4.1.

Main differences in the absorbance before (blue curve) and after UV curing (red curve) can be identified on the basis of three peaks in Figure 4.3. The key reaction of the C-double bond cracking is visible in the absorbance change at $1635.8\,cm^{-1}$ and $1616.5\,cm^{-1}$ labelled with indices 5 and 4, respectively. As listed in Table 4.1 the absolute intensities of these peaks are low compared to the other significant absorbance peaks (indices 1 and 3) and they appear in close vicinity to the before described C=O carbonyl peak. So strong absorption by its overtones in the NIR is not expected.

The most prominent signals of a changing absorption directly related to the chemical restructuring of the acrylate end group in Figure 4.3 are absorption peaks labelled with indices 1 and 3. These peaks are identified as the out of plane bending of $C-H$ at the acrylate end group $R-CH=CH_2$ at $808\,cm^{-1}$ and a scissor deformation of hydrogen atoms at the end group at $1406\,cm^{-1}$, respectively.

The normalized intensities of those two oscillations in the unpolymerized samples are more than three times as high compared to the C=C and $CH=CH_2$ absorbance signals. Hence the measurable difference between the polymerized and the unpolymerized state is highly significant.

Almost any publication examining photocurable acrylate coatings is using those two wavenumbers for the determination of the grade of acrylate conversation (e.g. Studer *et al.*, 2003 [14], Scherzer *et al.*, 2002 [44] or Yang *et al.*, 1993 [45]). In the following section 4.1.3 the most important overtones of these oscillations will be identified in the NIR.

The absorbance peak with index 2 at $983\,cm^{-1}$ is identified according to Hong *et al.*, 2005 [46] as the stretching vibration of the $C=C-C=O$ backbone. This oscillation also vanishes in the spectra of the polymerized sample compared to the spectra of the unpolymerized one due to the increased rigidity of the developed polymer network. Although this oscillation is not directly connected to chemical conversion of the acrylate end group, it is used for identifying the acrylate compound in unknown formulations.

In this context, all other not identified absorbance peaks in Figure 4.3 do not have a direct connection to the grade of polymerization, e.g. $C-O$ stretching vibration at around $1190\,cm^{-1}$ [44].

4.1.3   Overtone oscillation in the NIR

In Section 4.1.2 the main infrared absorptions were identified, which are correlated to the polymerization of the acrylate end group. These absorptions do have overtones in the near infrared and have been reported numerously in literature (Beuermann *et al.*, 1995 [49], Scherzer *et al.*, 2004 [50], Aldrige *et al.*, 1993 [51] or Urban *et al.*, 1995 [52]). The second overtone of the acrylate end group appears at a wavelength of $1620\,nm$ and is assigned to the $C-H$ stretching vibration of the acrylic double bond (see Figure 4.3 and index 1 in Table 4.1) according to Weyer, 1985 [53]. This overtone is

Index	Wavenumber	Normalized intensities	Assignment
1	808.9	0.65 (0.27)	Out of plane stretching of $C-H$ on $R-CH{=}CH_2$ [46, 47]
2	983.5	0.46 (0.13)	Stretching vibration of the $C{=}C-C{=}O$ [46]
3	1406.3	0.84 (0.28)	Scissor deformation of the $= CH_2$ [46, 44]
4	1616.5	0.23 (0.14)	Acryloyl double bond oscillation of the $C{=}C$ [48]
5	1635.8	0.23 (0.08)	Stretching vibration of the $CH{=}CH_2$ [46, 45]
6	1721.1	1.99 (1.98)	Stretching vibration of the acrylic carbonyl $C{=}O$ [45]

Table 4.1: Identification of the main absorbance peaks and their intensities of an unpolymerized (and polymerized) acrylate coating shown in Figure 4.3. The absorbance peaks with indices 1, 3, 4 and 5 are changing due to chemical reaction of the acrylate during polymerization. The absorbance with index 2 shows the rigidity increase of the polymer backbone. Index 6 is used for the normalization.

clearly isolated in the NIR spectra and is suitable for the monitoring of the conversion in UV cured acrylate coatings.

Figure 4.4 shows this absorption overtone in the unpolymerized sample (blue line). The polymerized sample does not show this absorption due to the complete polymerization. The change in absorption of this C-H oscillation is a direct measure of the degree of polymerization and accordingly an indicator for remaining acrylic C-double bonds in the coating. Furthermore an overall attenuation of the reflection is observed, as a result of the coating shrinkage in combination with a change of the refractive index ([51] and [54]) and a change in molecular scattering [55].

4.2 Acrylate coatings used in industry

Recently, the UV-coating industry has been developing photocuring formulations for new applications like corrosion protection in the automotive sector, electro insulation for the e-mobility lithium ion technology, lubricant varnishes for combustion engines or haptic coatings on packaging of consumer products. Several formulations, mostly based on acrylates were examined in the BMBF project. Often the photocuring principle is used in combination with the established thermal curing, due to the reliable and

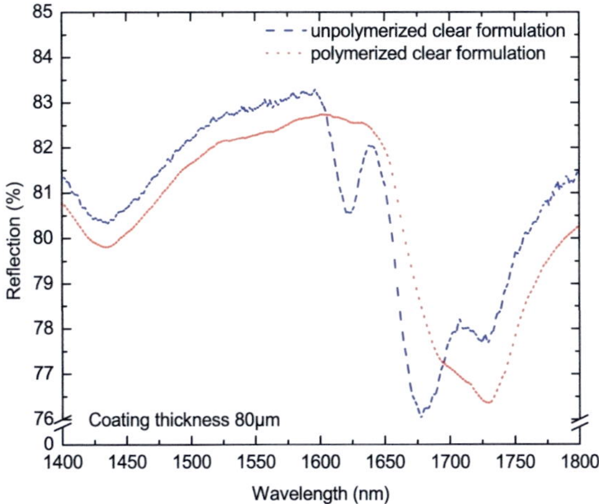

Figure 4.4: Near Infrared reflection spectra of the model formulation. The blue spectrum of the unpolymerized sample comprises the $1620\,nm$ overtone absorption, in contrast to the spectrum (red line) of the polymerized formulation. This overtone is associated to the $C-H$ stretching vibration of the acrylic double bond [53].

already established production facilities. Yet, exclusive application of the photocuring technology next to conventional solvent based inks is hardly found in the printing industry, as explained in Section 1.3.2. The following Section will give a rough overview of the possibilities of UV-curing and their applications, where formulations are under development or are already used in the production. All formulations are characterized by ATR-FTIR and analysed analogue to the formulation used in the experiments as discussed in Section 4.1.2.

4.2.1 Anti-corrosion coatings

Anti-corrosion coatings are characterized by coating thicknesses of $50\,\mu m$ to $80\,\mu m$ due to the necessity of being highly robust against water, salt and other environmental influences. The high coating thickness requires, for a thorough curing, a high amount of reactive compounds within the formulation. Figure 4.5 shows the ATR-FTIR spectra of a transparent anti corrosion coating used in the automotive industry. The coating was developed for the protection of aluminium housings comprising electronic control units for passenger safety and the brake system.

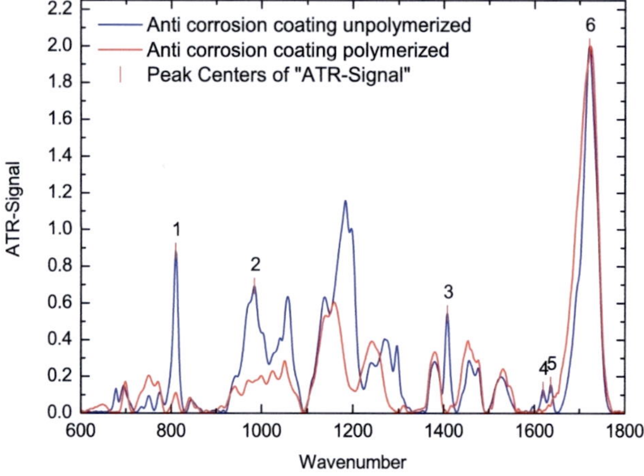

Figure 4.5: ATR-FTIR spectra of an anti corrosion protection formulation purely cured by photopolymerization. All main absorption peaks of an acrylate coating are evident and can be identified analogue to Figure 4.3. The labelled absorbance peaks are described in Table 4.2.

The results reveal a highly reactive acrylate coating, with a high share of C-double bond acrylate end groups. The similarities to the experimental formulation are remarkably. All characteristics of the experimental formulations can be found within this industrial formulation. See comparison of Table 4.2 and Table 4.1.

Index	Wavenumber	Intensity (normalized)
1	808.67	0.88
2	983.51	0.69
3	1406.49	0.54
4	1618.62	0.12
5	1635.33	0.15
6	1720.19	2

Table 4.2: Identification of the main absorbance peaks and their intensities of an unpolymerized (and polymerized) anti-corrosion coating shown in Figure 4.5. The absorbance peaks with indices 1, 3, 4 and 5 are decreasing with the acrylate polymerization. The absorbance with index 2 shows the rigidity increase of the polymer backbone. Absorbance with index 6 is used for normalization. (Note: The normalization is calculated to 2.)

This result impressively demonstrates the capabilities of even simple UV-coating formulations. They are able to fulfil such complex functionalities, like corrosion protection, with only few compounds. Especially, if the basic acrylate formulation is combined with anti-corrosion enhancers like nano-particles and other chemical compounds like oxidisers [56]. Commonly used solvent based formulations cannot be as easily modified nor dispersed.

4.2.2 Printing inks and top coatings

The main market share of the UV curing technology with 33% is the printing and packaging sector. Thus coating formulations used in this sector are most important. Offset printing and screen printing are mainly used for consumer goods like TetraPak$^©$ or tobacco packaging. The coating thicknesses are rather small, in the range of $1\,\mu m$ to $20\,\mu m$, compared to the anti-corrosion application. This property already implies some difficulties for the present sensor approach, discussed in Chapter 6. Still, the coating itself shows the same absorption peaks in the ATR-FTIR spectra like all other acrylate coatings, displayed in Figure 4.6.

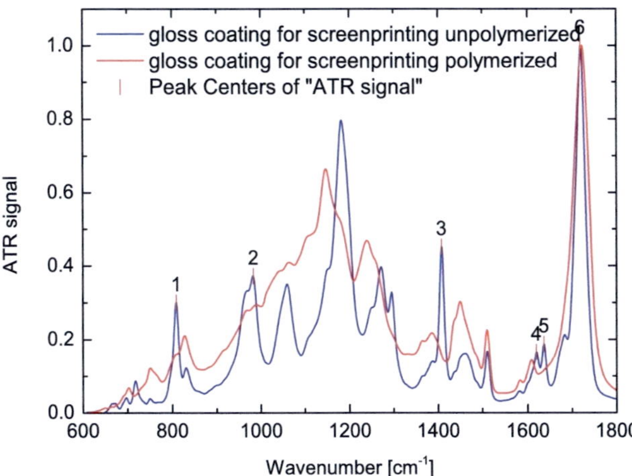

Figure 4.6: ATR-FTIR spectra of a gloss coat for printing applications. All acrylate related absorption peaks are present and identified according to Table 4.1 and summarized in Table 4.3. The absorbance peaks with indices 1, 3, 4 and 5 are decreasing with the acrylate polymerization. The absorbance with index 2 shows the rigidity increase of the polymer backbone. Absorbance with index 6 is used for normalization.

The absorption peaks are summarized in Table 4.3 accordingly to the results of the basic model formulation discussed in Section 4.1.2. The examined coating is used for finishing the coating of a consumer good package to obtain a gloss effect.

Index	Wavenumber	Intensity (normalized)
1	809.34	0.29
2	982.23	0.37
3	1406.49	0.45
4	1618.62	0.16
5	1635.33	0.18
6	1718.90	1

Table 4.3: Summary of the absorbance peaks, identified in Figure 4.6, of the gloss coating used in the packaging industry.

The high absorbance of the C-H ligands (indices 1 and 3) will be dramatically reduced, if the coating is applied as thin top coat and the overtone oscillation is monitored. The following Section 4.3 discusses this issue in more detail on a complex two-component formulation, which has to be thermally dried as well.

4.3   Two-component, solvent-based formulations

Today's most spread coatings are solvent-based and form two-component systems and their combinations. Those systems are highly complex, due to the many compounds necessary for a sufficient cross linking and the resulting functionality. An upcoming application of solvent-based two-component formulations are coatings for tribology improvement, known as low friction coatings and electrical insulating systems for high voltage application in harsh environments. Application areas are mainly in combustion engines as well as in joints, tools or mechanical gear systems and actuators used for fuel injection. Due to the complexity, a sufficient quality monitoring tool for these systems does not yet exist. Attempts to develop such a tool are within the scope of the underlying BMBF project of this work.

Figure 4.7 shows the ATR-FTIR spectra of a highly efficient electrical insulating coating used in the automotive sector. The complete polymerization of the formulation is essential for the functionality of the insulation. The polymerization process (polyaddition) is performed within two steps. First the coating is UV irradiated and afterwards hot dried for sev-

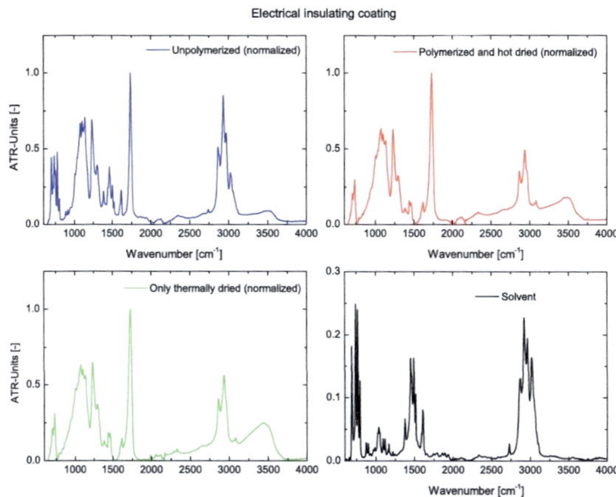

Figure 4.7: ATR-FTIR spectra from an electrical insulating system. Top left shows the complete unpolymerized formulation including the solvent. The correct polymerized and thermally dried coating is displayed in the top right. The spectral data of a sample only dried thermally is shown in the bottom left corner. The pure solvent absorption spectrum is displayed at the bottom, right.

eral hours. Figure 4.7 shows an overview on these steps, unpolymerized solvent-containing formulation (top left), polymerized and hot dried formulation (top right), solely hot dried (bottom left) and the pure solvent (bottom right).

The spectra does not clearly show any of the above discussed absorption peaks typical for an acrylate. The main composites of the formulations are polyester and melamine resin. The main differences in the ATR-FTIR spectra (Figure 4.8) are due to the evaporation of the solvent. The differences of the polymerized and thermally dried formulations (bottom left) are more than one magnitude smaller than to the differences between the unpolymerized and thermally dried (top right) and the completely polymerized and unpolymerized (top right) formulations. The most obvious change in the ATR-FTIR spectra is due to the solvent content. Thus the grade of conversion is not necessarily comprised in this result. In order to measure the grade of conversion using ATR-FTIR detailed informations on the chemistry are lacking.

The FTIR-ATR information given in Figure 4.7 and the difference plots in Figure 4.8 indicate, due to absorbance changes in the range of $800\,cm^{-1}$

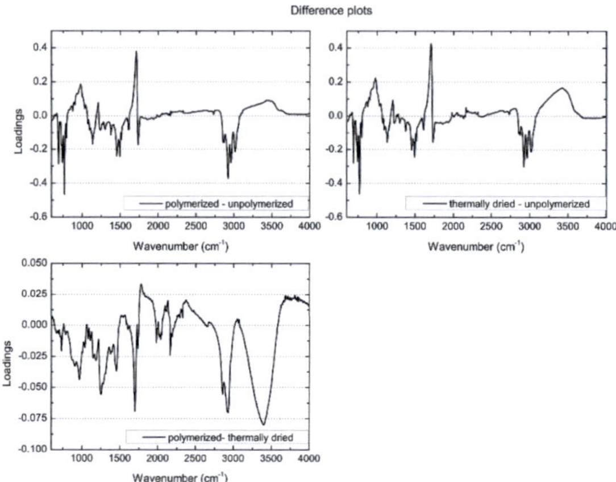

Figure 4.8: The ATR-FTIR absorbance differences between the unpoly-merized and the polymerized samples are shown in the top left graph. Main differences with amplitude of about 0.35 are at $900\,cm^{-1}$ and $2800\,cm^{-1}$. The difference plot of thermally dried and unpolymerized samples (top right) shows approximately the same spectral absorption differences and amplitudes as the difference between raw and hot dried, polymerized coating. On the left bottom side the differences of the thermally dried and polymerized sample is shown. The amplitude is only one tenth of the amplitude observed in the other two spectra. Therefore only the solvent does give a proper spectral absorption response in the ATR-FTIR signal.

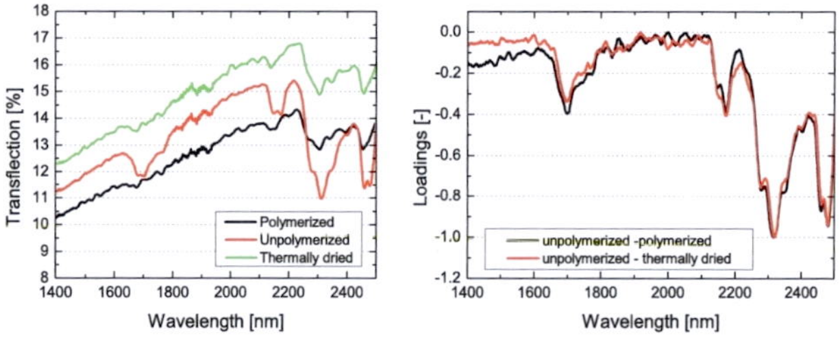

Figure 4.9: NIR spectra and differences of the electro insulation coating. The left graph shows the spectra of the three possible physical conditions of the coating. Unpolymerized, polymerized and only thermally dried. The main difference in the absorption at $1680\,nm$,$2150\,nm$ and $2300\,nm$ is due to the evaporation of the solvent. The differences plot on the left reveal this dependence only on the solvent more clearly.

to $1000\,cm^{-1}$, possible overtone oscillations in the NIR analogue to the acrylate behaviour. Yet, as shown in Figure 4.9 there is no evidence in the subtracted NIR-spectra in the range from $1400\,nm$ to $2400\,nm$. The left graph shows the unpolymerized and hot dried, polymerized formulations as well as the pure solvent. The calculated differences between the unpolymerized and polymerized formulations and the polymerized data minus the data of the thermally dried sample is shown on the right.

The principle observations in the Mid Infrared can be copied to the Near Infrared. The main absorption changes in the spectra are caused by the evaporation of the solvent. The actual polymerization process is not clearly observed in this spectral range. Figure 4.9 (left) only shows absorption differences at about $1680\,nm$, $2150\,nm$ and most prominent at $2300\,nm$ due to the thermal drying. Again a conversion monitoring dependent on the actual amount of polymerized molecules is not possible.

Chapter 5

Measurement principle and hardware design

This chapter discusses the main principle used for the detection of the acrylic conversion without knowing the complete NIR spectra. The highly efficient transflection principle used in this sensor system and its normalization procedure is explained. The information needed, can be acquired by only using two laser sources selected for this task. Most important for a reliable detection of the acrylate conversion, after or while the photo-curing, is the ability to excerpt the desired transflected signal, which is only dependent on the chemical properties of the coating.

Furthermore the development process is always based on a broad fundamental research of all necessary hardware components. The hardware part of the developed sensor system can be divided into several conceptional tasks:

- Optical concept for the contact free measurement
- Suitable radiation sources and detectors
- Signal acquisition and processing
- Provisions for low noise
- Laboratory set-up of the complete system

These tasks will be discussed in the following chapter. The main optical concept for the contact free measurement is most important and has not yet been realized before. In the first Section of the chapter, two different possible solutions are compared and illustrated. The decision for the laser sources and the detectors is explained in Section 5.1. Due to the absorption spectra of the target coating (see Section 4.1.3) and the precondition of being cost effective, the possible solutions are already restrained. Suitable hardware for the signal acquisition and processing is closely connected to the used detector and discussed afterwards. The aspect of a high signal to noise ratio for a reliable signal evaluation is as important as having an effective optical concept. The used Lock-In technique is addressed in Section 5.3.

The measurement set-up and the different steps in optimizing, leading to the final laboratory based sensor system is shown in the last section.

5.1   Sources and Detectors

The Near-Infrared region from $800\,nm$ to about $1800\,nm$ is not detectable for the most common detectors made of silicon. The band gap of silicon has a value of $\approx 1.1\,eV$; i.e. the wavelength limit for the electron transfer from the valence to the conduction band exists at $1127\,nm$ in vacuum. Practically this limit is even lower and most silicon detectors stop detection at less than $1000\,nm$. The material of choice for the above named wavelength region is Indium doped Gallium-Arsenide (InGaAs) with a band gap of pure Gallium-Arsenide of $1.42\,eV$ and Indium-Arsenide of $0.36\,eV$. Depending on the proportion of the two materials the InGaAs semiconductor above $\approx 0.98\,eV$ exhibit a direct band gap and below an indirect band gap (for more details see Sze *et al.*, 2007 [57]).

InGaAs semiconductor material can be used both as a laser and detector; only the manufacturing process is different. The sources used in the sensor are twofold: The main wavelength at $1620\,nm$ responsible for the polymer conversion detection is provided by a highly precise DFB-laser to ensure a sufficient accuracy. The normalization wavelength at $1550\,nm$ is provided by a simpler and cheaper Fabry-Pérot laser diode. The Fabry-Pérot laser diode used in the laboratory set-up was purchased from Mitsubishi Electric (see Table 5.1 for more details).

Centre wavelength (nm)			Spectral Width (nm)			Optical Output (mW)
min.	typ.	max.	min.	typ.	max.	typ.
1520	1550	1570	-	1.5	3	6

Table 5.1: Specifications of the Fabry-Pérot laser diode used as reference in the sensor system.

The detection of both signals is provided by a cooled InGaAs detector operated at $-40°C$. The DFB-laser diode principle and the responsivity of the InGaAs detector will be described in the following paragraphs.

5.1.1 Distributed Feedback (DFB) Laser

Distributed Feedback Lasers are characterized by precise wavelength selectivity with a FWHM in the order of less than one nanometre. Additionally the DFB laser exhibits an excellent higher mode damping of about $50\,dB$. In contrast to the common Fabry-Pérot laser diode the DFB diode with lateral coupling is built of the active semi-conductive laser material and a refractive index modulated grating is responsible for the mode selection within the device. Figure 5.1 (a) shows a typical set-up of such a DFB laser unit. The high wavelength selectivity of a DFB laser is achieved by modulating the pitch of the refractive index change within the device. The desired wavelength can be described by using the Bragg condition:

$$\lambda = 2 \cdot n_{eff} \cdot d \tag{5.1}$$

with λ for the emitted wavelength, n_{eff} the effective refractive index and d the spacing of the grating. The effective refractive index n_{eff} can be further described by

$$n_{eff} = \frac{\beta}{k_0} = \beta \cdot \frac{\lambda}{2\pi} \tag{5.2}$$

with β as the propagation constant and k_0 as free space wavenumber [58].

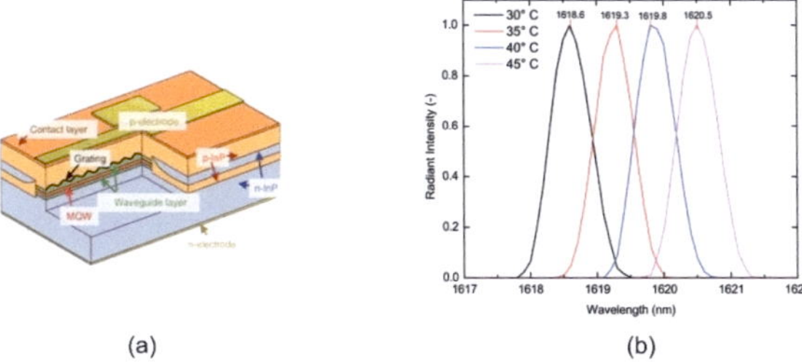

Figure 5.1: (a): DFB Laser principle (Image source: [59]) (b): Sacher DFB Laser emission, depending on temperature with nominal wavelength of $1620\,nm$ used in the experiments. Optical output power according to manufacturer: 5 mW (cw). Peaks of emission are added in the graph.

Figure 5.1 (b) shows the temperature dependence of emission due to the thermal expansion of the grating. The used DFB laser diode is, due to this effect, thermally stabilized at 45°C using a laser driver from Sacher Lasertechnik Group. The feasible wavelength shift by controlling the temperature is in the range of some nanometres. The used device can be tuned by $0.125\,nm$ per degree Celsius. DFB laser development in the NIR and examples for its application can be found in Sato *et.al.*, 2009, Koeth, 2008 or Zeller *et.al.*, 2010 [59, 60, 61].

Centre wavelength (nm)	Spectral Width (nm)	Optical Output (mW)
1619.8	± 0.28	5

Table 5.2: Specifications of the DFB laser diode used for the detection of the conversion in the sensor system. Spectral width acquired by a Gaussian fit with the spectral data from Figure 5.1 (b) at 40 °C.

5.1.2   NIR detector: Indium-Gallium-Arsenide

The detector was purchased from Teledyne Judson Technologies with the specifications summarized in Table 5.3. The active sensor area is $7.06\,mm^2$ with a diameter of $3\,mm$. The spectral response curve and the principal design are shown in Figure 5.2. The detector is cooled with a two step thermoelectric element to -40 °C to achieve low values of noise equivalent power (NEP).

The signal output voltage of the active sensor area is enhanced by an adjustable factor of 10^4, 10^5 or 10^6, respectively. Afterwards the signal is fed into a high precision voltmeter or Lock-In amplifier and processed by the sensor's LabView software.

The broad spectral response of the InGaAs diode was chosen on early stage project considerations to cover a wide spectral range within the near infrared region. A more effective diode tailored to the $1620\,nm$ region is used in the final prototype.

5.1.3   UV radiation source for in situ polymerization measurement

Common UV radiation sources are medium pressure lamps with some Kilowatts of power consumption and an elongated lamp shape, as shown in Figure 2.7. This type of radiation source is not suited for spot irradiation of

Peak Response (A/W) @ -40°C	1.28
Shunt Impedance (Ω)	155k
Detectability ($cm \cdot \sqrt{Hz}/W$)	11.8 E+11
Thermistor resistance (Ω) @ -40°C	19k
Cooler Current (A) @ -40°C	0.78
Active size diameter (mm)	3
50% Cut-off Wavelength (μm)	2.31 $\pm$ 0.1
NEP ($W/\sqrt{Hz}$) @ peak wavelength (typ.)	2.80 E-13
Capacitance (pF) @ 0V (typ.)	9000

Table 5.3: Specification for the used InGaAs detector from Teledyne Judson Technologies according to the individual technical data sheet. For comparison the NEP of the same detector operated at -20°C is typically $5.9^{-13}\,W/\sqrt{Hz}$.

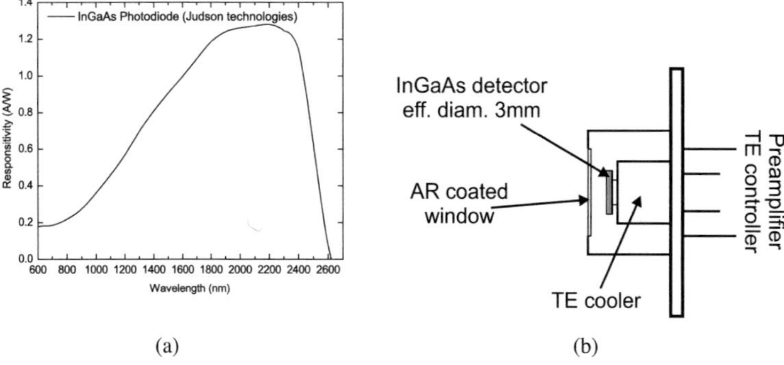

(a) (b)

Figure 5.2: Spectral response and technical details of the InGaAs detector used in the laboratory set-up. (a) Spectral response curve of the J23TE2-66C-Ro3M-2.4 InGaAs detector as purchased from Teledyne Judson Technology. (b) Technical drawing of that detector. (All data provided by Teledyne Judson Technology.)

acrylates to be polymerized and an in situ monitoring of the acrylic polymerization process, later described in Chapter 6. A better solution is provided by high pressure argon lamps (Figure 5.3). These lamps can be easily focused and coupled into light guides for a maximum of flexibility in the applications, like UV-adhesives or contact lens manufacturing.

The light guide with a diameter of 8 mm is mounted onto the front part of the spatial mirror cage and irradiates the sample with an angle of incidence of 30°. The irradiance of the UV radiation source is measured in depen-

Figure 5.3: The HPA 250-500 produced by Philips is used in the irradiation system from Dr. Gröbel UV Elektronik GmbH.

Equation	a	b
$y = a \cdot x^b$	8304.20 ± 500.78	-1.76 ± 0.07
	Reduced Chi-square	corrected R^2
	2165.19	0.998
UV-A	UV-B	UV-C
40.0%	14.2%	4.4%
	UV	VIS
	58.6%	41.4%

Table 5.4: UV source power distribution and fit parameter

dence of the distance from the light guide surface by a spectroradiometer calibrated to a PTB standard [62]. The radiation system delivers mainly UV-A (40%) and visible radiation (41.4%). UV-C and UV-B is, due to the Argon spectrum less available and its proportion is 4.4% and 14.2%, respectively (see Table 5.4). The irradiance measurement is shown in Figure 5.4. The decrease of the irradiance can be fitted by a simple non-linear function $f(x) = a \cdot x^b$, which complies to the assumption of an almost quadratic decrease of the irradiation with increasing distance. The near field approximation is defined by $b = -1$ and the far field approximation to $b = -2$.

The actual irradiance in the laboratory set-up on the sample under an angle of 30° and a distance of 5 cm is measured to $I_{30°} = 311.8\,W/m^2$ in the range of 200 to 780 nm. The shutter of the UV radiation system is connected to the LabView control unit and can be continuously adjusted from 100 ms to several seconds of exposure time, as well as the electrical power from 100% down to 20%.

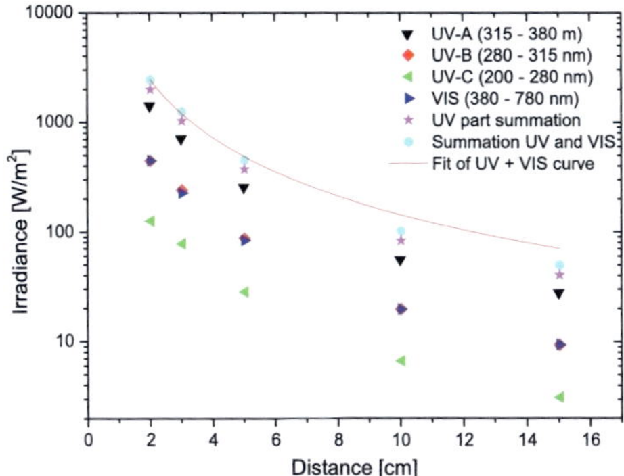

Figure 5.4: Irradiance measurement on the UV radiation system LQ-400 purchased from Dr. Gröbel UV Elektronik GmbH, Germany. The high-pressure radiation source is filled with Argon and has a rated electric power consumption of $400\ W$. The irradiance was measured with a spectroradiometer calibrated to a reference radiation sources (Deuterium and Halogen lamp) from the PTB Germany.

5.2 Implementation of the transflection concept in the sensor system

The realization of a contact free optical measurement system encounters two principal problems. First, the radiation, which is guided onto the sample, transflected from there and received at the detector, interferes with itself during this process. For addressing this problem, two solutions are considered. A beam splitter or a spatial filter can be used for that. Second, the incident light beam has to be kept perpendicular to the reflection surface, in order to maximize the detection signal (see Chapter 6, Section 6.5). This positioning is treated as a boundary condition and has to be fulfilled within certain limits while installing the sensor system. Therefore the measurement will be always performed with the incident light being perpendicular to the sample surface.

Figure 5.5(a) shows the schematic set-up of all components in the laboratory demonstrator. The two laser sources are combined into an identical optical path. The light chopper modulates both beams for the Lock-In amplifier processing of the received signal.

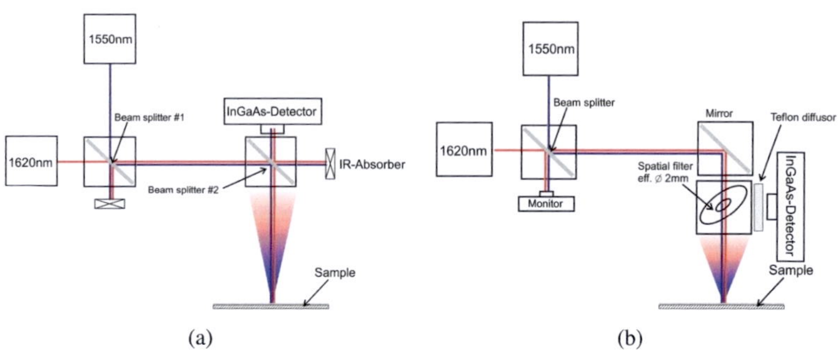

Figure 5.5: (a) Laboratory set-up with two beam splitters. (b) Laboratory set-up with one beam splitter and a spatial mirror filter. The two laser sources are combined to propagate on identical paths by a beam splitter.

5.2.1   Concept with two beam splitters

In Figure 5.5(a) the first laboratory set-up for the conductance of the laser light with two beam splitters is shown.

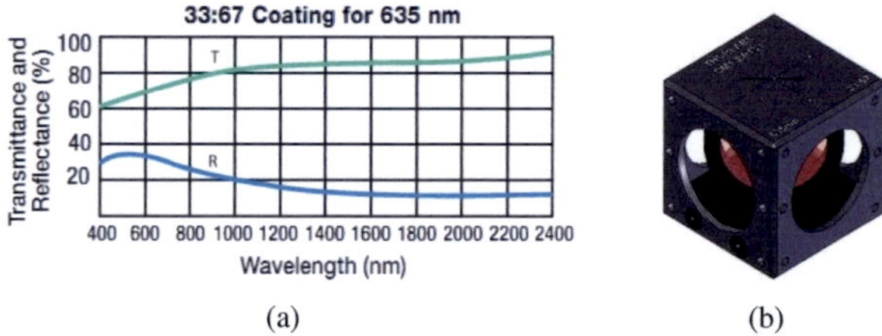

Figure 5.6: Beam splitter used in the laboratory set-up, purchased from Thorlabs Inc. (Cube-Mounted Pellicle Beam splitter, 33:67, 635 nm). a) Spectral Transmission and Reflection according to Thorlabs. b) The mounted beam splitter in a 30x30 mm lab cube.

The laser light is combined by using a first beam splitter described in Figure 5.6. This beam splitter combines the light path of the laser sources. After passing the chopper used for the Lock-In amplifier (see section 5.3), the second beam splitter is used for the collection of the transflected radiation scattered back from the sample surface. The optical properties of the beam splitter are optimized at a wavelength of $635\,nm$. The ratio of reflectance to transmittance is given to 33:67 at the optimized wavelength. In the NIR range this ratio changes to a transmission of 85% and a reflection of 15%, respectively. The maximum signal amplitude I of each laser at the detector can be estimated :

$$
\begin{aligned}
LS: \quad & I_{1550nm} = 1.0 & I_{1620\ nm} &= 1.0 \\
BS1: \quad & I_{1550nm} = 0.15 \cdot 1.0 & I_{1620nm} &= 0.85 \cdot 1.0 \\
BS2.1: \quad & I_{1550nm} = 0.15 \cdot 0.15 \cdot 1.0 & I_{1620\ nm} &= 0.15 \cdot 0.85 \cdot 1.0 \\
BS2.2: \quad & I_{1550nm} = 0.85 \cdot 0.15 \cdot 0.15 \cdot 1.0 & I_{1620nm} &= 0.85 \cdot 0.15 \cdot 0.85 \cdot 1.0 \\
DT: \quad & I_{1550nm} = 0.019 & I_{1620nm} &= 0.108
\end{aligned}
\tag{5.3}
$$

The maximum receivable laser radiation amplitude for this two-beam splitter approach is approximately a tenth of the original radiation amplitude of the $1620\,nm$ laser source. Due to the two reflections of the $1550\,nm$ laser source, the amplitude $I_{1550\,nm}$ even drops to about 2%.
This weak exploit of laser radiation has led to the development of a more suitable optical concept, including a spatial filter mirror instead of a second beam splitter (see next Section 5.2.2).

5.2.2 Spatial filter approach

The use of a spatial filter, instead of a beam splitter seems to be the most promising solution in this kind of optical set-up of the sensor system. Equation 5.4 calculates the maximum relative detection amplitude for this set-up. Compared to the results of Equation 5.3 the signal gain is one magnitude for the $1550\,nm$ signal and about eightfold for the $1620\,nm$ laser source.

$$
\begin{aligned}
LS: \quad & I_{1550\ nm} = 1.0 & I_{1620\ nm} &= 1.0 \\
BS1: \quad & I_{1550\ nm} = 0.15 \cdot 1.0 & I_{1620\ nm} &= 0.85 \cdot 1.0 \\
SF: \quad & I_{1550\ nm} = 1.0 \cdot 0.15 \cdot 1.0 & I_{1620\ nm} &= 1.0 \cdot 0.85 \cdot 1.0 \\
DT: \quad & I_{1550\ nm} = 0.15 & I_{1620\ nm} &= 0.85
\end{aligned}
\tag{5.4}
$$

The spatial filter is made of an infrared optimized gold coated mirror with an elongated centre aperture for the incoming laser light. The aperture is calculated to an effective radius of $2.5\,mm$ with respect to the mounted angle of $45°$. Figure 5.7(a) shows the spatial filter and Figure 5.7(b) the mounting position. The transflected laser light from the sample is almost completely reflected by the gold mirror. The spatial filter has the outer diameter of $25.4\,mm$ ($1''$) and the aperture measures $4\,x\,9\,mm$. The loss of radiation due to the aperture can be calculated to 6.4% in relation to the overall mirror surface.

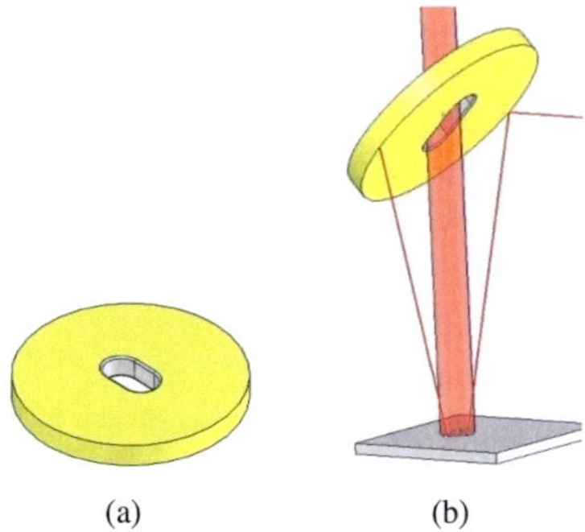

(a)   (b)

Figure 5.7: (a): The spatial filter as designed for the sensor system (CAD drawing). (b): The spatial filter virtually mounted in the sensor system. The angle of the mirror is adjusted to $45°$. The laser light is adjusted to fit through the aperture and the reflected radiation is guided via the gold surface to the detector unit (not displayed).

Further optimizations by focusing and optical diffusion discs will be discussed later in this chapter. The above discussed issues of the spatial filter approach and its design have been developed by Elmar Feuerbacher[1] in close cooperation with the CuringOnlineSensor team, with special mention of and Oliver Treichel[2].

5.3 Raising the signal to noise ratio: Lock-In Amplifier

The raw detector signal without any further processing is strongly influenced by the detector and laser noise. Furthermore a reliable measurement of different laser signals at once is only possible by modulating each signal at a non interfering frequency with respect to the other signals. The laboratory set-up has to be already prepared for this option in later development stages (see Chapter 7). In the laboratory demonstrator, the signals are sequentially processed at a global chopper frequency of $1\,kHz$.

The signal to noise ratio is improved by the Lock-In from $100 : 5$ up to $1000 : 1$ with an integration time of $100\,ms$. The processed signals of the

[1]QIAGEN Lake Constance GMBH, Elmar.Feuerbacher@qiagen.com
[2]formerly IST Metz GmbH

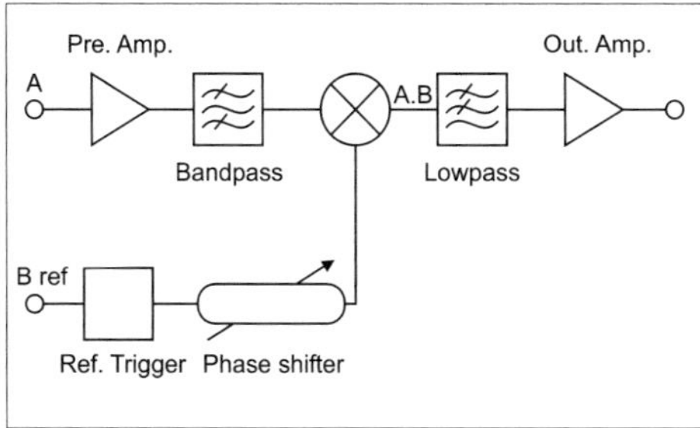

Figure 5.8: Lock-In technique used for the improvement of the signal to noise ratio. The laser signals are amplitude modulated by an optical chopper at a frequency of some kHz. The reference signal and the detector signal are multiplied and low pass filtered, which corresponds to a Fast-Fourier Transformation. The phase of the laser radiation has to be taken into account each time the laser is restarted to acquire a constant and maximum amplitude.

$1550\,nm$ and the $1620\,nm$ lasers are then used to compute the conversion grade of the polymer (see section 5.4 and equation 5.5).
The measurements are conducted with a digital Lock-In (Model 7265) amplifier from Signal Recovery [63]. This frequency range is from $0.001\,Hz$ to $250\,kHz$ with a voltage range from $2\,nV$ to $1\,V$. The phase noise is defined to $< 0.0001°$ rms.

5.4 Implementation and processing of the discrete NIR-laser signals

According to the NIR-spectra measurements (see Section 4.1.3), there is a need for referencing the conversion signal at $1620\,nm$ to an independent NIR-reflection signal, due to the change of the overall reflection caused by the change of the refractive index and scattering behaviour. The reference laser with emission at $1550\,nm$ was chosen for several reasons: The reference wavelength has to be close enough to the actual $1620\,nm$ absorption to be able to assume a linear spectrum in between the two frequencies. Furthermore the reference source must have a higher or similar radiation power and should be low priced as well as always available. Early experiments with laser sources radiating around $900\,nm$ to $1000\,nm$ showed a reduced performance in the linear spectrum assumption and thus high deviations in the calculated signal quotient and so have not been further investigated.

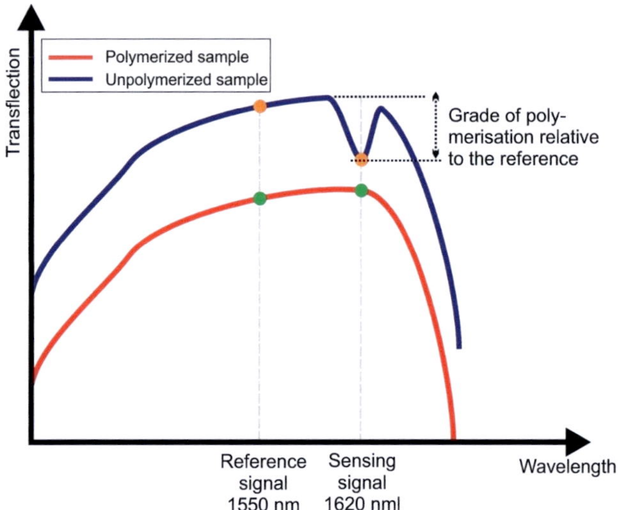

Figure 5.9: Principle of the data acquisition using only two independent signals from the spectrum. The normalization at $1550\,nm$ and the actual conversion at $1620\,nm$. The conversion is computed by comparing the actual quotient C with a polymerized value.

Figure 5.9 shows the main principle of referencing the spectra described before in Figure 4.4. The correlated transflection signal of the acrylate overtone of the $C-H$ stretching oscillation as well as the transflection independent from polymerization are measured. These two values are used for the determination of the conversion grade of the polymer as follows:

$$C_{acrylate} = \frac{I_{1620\,nm}}{I_{1550\,nm}} \tag{5.5}$$

with $I_{1620\,nm}$, $I_{1550\,nm}$ and C $(0 \leq C \geq 1)$ as the intensities of the transflected laser radiations and the relative conversion, respectively. The comparison of the conversion quotient C of the actual sample with the conversion value C^* of a completely polymerized formulation results in the percentage of polymerization C' :

$$C' = \frac{C}{C^*} \cdot 100[\%] \tag{5.6}$$

The above described percentage of polymerization should not be set equal with the often used degree of conversion in literature. The degree of conversion of the acrylate is calculated according to Decker *et al.*, 1988 [64] from the FTIR data as follows:

$$\text{Degree of conversion} \quad C[\%] = \frac{(A_\Lambda)_0 - (A_\Lambda)_t}{(A_\Lambda)_0} \cdot 100\% \qquad (5.7)$$

with A_Λ for the correlated absorbance by the fundamental oscillations of the acrylate system in the far infrared. The initial unpolymerized state is indicated by $(A_\Lambda)_0$ and the all discrete time steps during the polymerization are indicated by $(A_\Lambda)_t$. The degree of conversion calculated from FTIR spectra becomes equal 100% if the absorption completely disappears. However a complete polymerization is not possible. Thus a value of 100% is only of theoretically nature. Typical values achieved in laboratory experiments are above 70%, depending on the UV radiation source, atmospheric conditions and used photoinitiators [15, 65, 50].

This simple but robust approach is capable of detecting the acrylic conversion in simple coating systems like the model formulation described in Section 4.1.1. The final sensor system will be using two different referencing wavelengths, due to several constraints resulting from the more complex industrial formulations (see Chapter 7).

5.5   Sensor set-up in the laboratory

The transflective measurement principle (Section 1.2) and the above described (Section 5.4) spectral acquisition of the conversion information has been realized in a laboratory set-up as shown in Figure 5.12. The set-up is designed to be most flexible with all optical components, radiation sources, detectors and data processing including Lock-In amplifier and modular LabView programming. The laser sources are combined via a beam splitter (Figure 5.6), instead of using a fibre coupling as done later during the development process (Chapter 7). The laser sources are operated in CW-mode and are modulated for the Lock-In amplifier by an optical chopper wheel. Therefore the measurement of each laser signal has to be done sequentially on the same sample path and the data are combined afterwards to determine the grade of conversion.

The complete set-up in Figure 5.12 is made with standard optical components (30 mm cage system) purchased from Thorlabs. The operation mode of these components is shown in Figure 5.10. The transflected laser light is detected by the InGaAs diode and processed by a pre-amplifier and the Lock-In amplifier and eventually recorded and graphically displayed in the

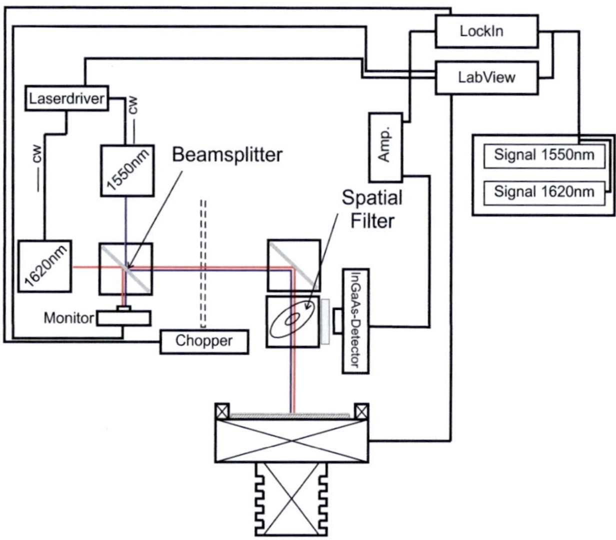

Figure 5.10: Principle of the data acquisition as realized in the laboratory. The two laser sources are coupled via a beam splitter and guided trough the spatial filer onto the sample. The stage is movable measuring along a variable linear distance on the samples. The signals are detected with the InGaAs module and processed using LabView.

LabView software unit. The best signal stability has been achieved with a Teflon diffuser in front of the detector to ensure a uniformly distributed irradiation of the InGaAs chip. The movable stage is driven by a stepper motor and controlled by the LabView software unit. The linear stage LS 110 and elevation stage ES 100, purchased from MICOS GmbH, are combined and movable in height and lateral direction with a repeated accuracy of $\pm\,1\mu m$ at all travel distances (max. $305\,mm$) and moving speeds up to $50\,mm/sec$. The sample is placed on the stage and magnetically locked. To verify a vertical alignment of the lasers with respect to the sample stage, a mirror is used to adjust the optical path. The best alignment result is achieved as soon as a global signal minimum on the detector is recorded. This means nearly the entire laser light is directly reflected back trough the pinhole of the mirror, in case the angle of incidence is kept to $90°$.

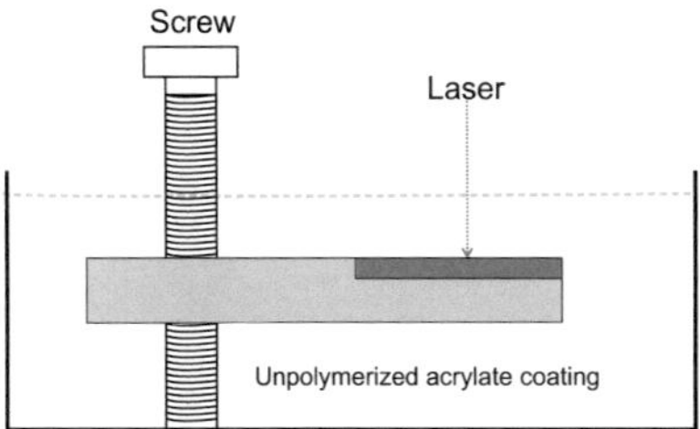

Figure 5.11: Illustration of the measuring device used for depth profiling of the clear and pigmented formulations. The tray is filled with the liquid formulation up to a defined mark. By turning the screw a substrate is moved vertically through the formulation, thus decreasing the coating thickness from $4000\,\mu m$ down to nearly zero. Unintended volume changes in the tray are avoided in this configuration and can be neglected.

5.5.1 Depth profiling tool for clear and pigmented unpolymerized acrylate formulations

The penetration depth of the sensor system into the acrylate system is measured by using a substrate probe mounted on a boom. The boom is moved stepwise through the acrylate resin filled tray up to the surface of the liquid coating, see Figure 5.11. The set-up ensures a constant resin volume in the tray to minimize measurement errors of the layer thickness. The set-up adjusts the film thickness with a resolution of $85\,\mu m$ per step.

The depth profiling can only be performed with unpolymerized liquid samples, due to the necessary movement of the boom. The tool was proved by an empty tray measurement for a constant vertical laser beam adjustment throughout the complete 40 mm travel distance. The experimental results acquired by this tool are discussed in Section 6.3 in detail. The experimental result acquired with the new developed and here described method allows the determination of the actual information depth of the sensor as a function of coating thickness and pigment concentration.

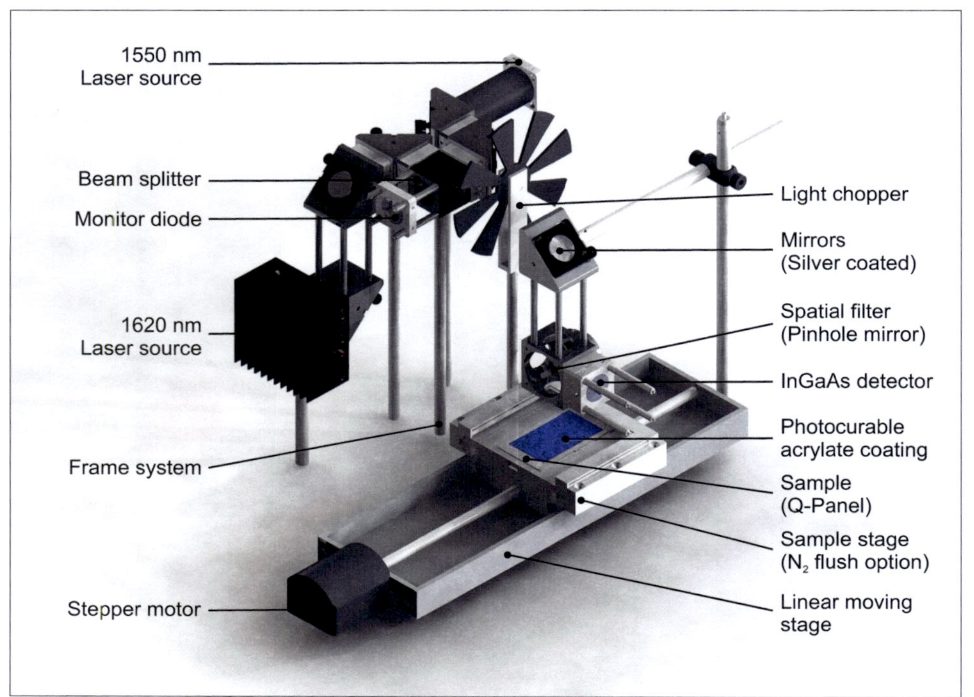

Figure 5.12: The used experimental set-up as a CAD drawing. All components are designed to be replaceable and adjustable to ensure a most flexible development environment.

Chapter 6

Experimental results of the discrete laser system

The feasibility of a discrete laser system capable of monitoring the acrylate conversion in many UV curable coatings and the development of the laboratory based demonstrator is presented in this chapter. In Section 6.1 the main principle of detection is proved by measurements on a model formulation consisting of the main components of a UV formulation (see Section 4.1).

The difference signal from an unpolymerized and polymerized sample is recorded as well as the progression while being irradiated by an UV source. A cross-check and validation of the results are realized by measuring the same samples with the Lambda1050 (see Section 3.4.2). Thus it is possible to reconstruct the sensor signals in a calibrated device.

Section 6.2 presents the principal boundaries of that approach neglecting other limitations like tailored industry formulations or system related noise. Different coating thicknesses and their signal differences are used to determine the minimum layer thickness for a significant conversion detection. Another very important part covers the analysis of the maximum penetration depth of the laser into a coating, resulting in the maximum information depth.

Section 6.3 describes measurements in clear and pigmented coatings to get most relevant data with respect to industry needs. Part 6.4 presents conversion data for different irradiation times to reveal possible savings in time and thus production costs.

6.1 Optical signal vs. C-double bond conversion in acrylate systems

Figure 6.1 shows a typical polymerization cycle of TPGDA with Irgacure 819 as photoinitiator. TPGDA (see Figure 2.2) is a simple acrylic high reactive monomer. The sample consists of a droplet placed on the Q-Panel substrate to achieve a high coating thickness. The irradiation time was set to two seconds to ensure a maximum of C-double bond conversion. The $1620\,nm$ laser signal increases immediately after illumination from a value of 0.87 ± 0.022 to the final value of 1.00 ± 0.004 after the UV radiation was shut off again. The $1550\,nm$ laser signal does not change significantly during the polymerization process. The slight increase of fluctuations in the $1550\,nm$ signal, visible right after opening and closing the shutter, arises from the abrupt change in the illumination condition of the detector. The high fluctuations in the $1620\,nm$ signal originate from the surface vibration of the TPGDA droplet, caused by the nitrogen flow used during polymerization and the low viscosity of TPGDA.

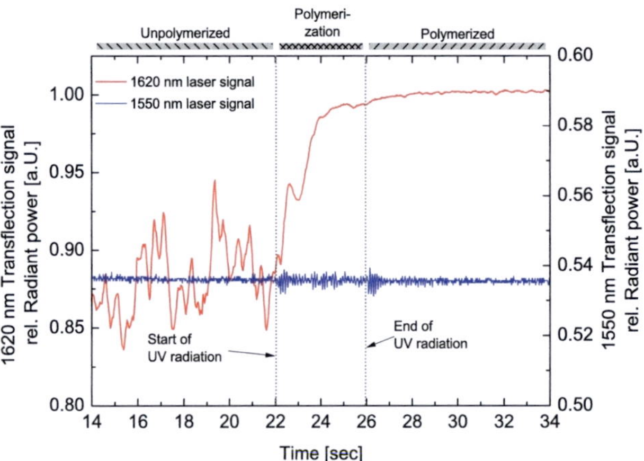

Figure 6.1:   Time resolved polymerization of Tri(propylene glycol)diacrylate (TPGDA). The red line reflects the C-double bond overtone absorption at $1620\,nm$ (left ordinate). The blue signal shows the reference signal at $1550\,nm$ (right ordinate). The start and end of UV illumination is displayed and labelled.

The sensor system is set up as previously described and all measurements were performed with constant laser powers and detector adjustments. The remarkably constant $1550\,nm$ laser signal shows, that the optical properties of pure TPGDA do not change in the Infrared, except for the C-double bond overtone at $1620\,nm$. This effect is clearly observed only in this particular simple case. More complex formulations used for the experiments later will prove the need for at least one reference wavelength.

The amplitude change observed in Figure 6.1 is more than 10%. TPGDA contains a high C-double bond concentration due to a low molar mass of $\approx 300\,g/mol$ compared to the later used modified epoxyacrylate with a higher molar mass of $\approx 580\,g/mol$. Thus pure TPGDA will always show the highest signal change with respect to the same coating thickness.

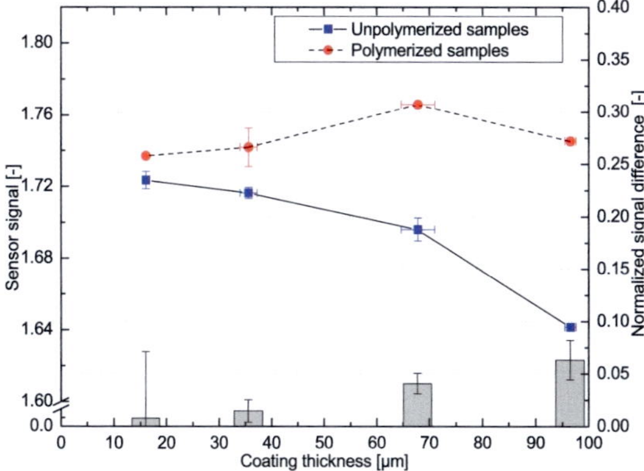

Figure 6.2: Experimental signal result from unpolymerized and polymerized samples acquired by the discrete laser spectroscopic system. Left ordinate shows the absolute radiant quotient of the sensor system. The bar chart (right ordinate) shows the signal difference of the coatings before and after UV curing. The error bars are the standard deviation of five independent samples. The coating thickness of each measurement was determined after curing.

Figure 6.2 shows the quotient $C_{acrylate}$ (see equation 5.5) for the standard model formulation from polymerized and unpolymerized samples as a function of coating thickness. Samples coated with $16\,\mu m$, $35\,\mu m$, $67\,\mu m$ and $96\,\mu m$ clear formulations respectively, were measured before and after UV curing using the above described experimental set-up. The radiant quotients of signal intensities are plotted in Figure 6.2, as well as the radiant quotient differences between unpolymerized and polymerized samples normalized with $y' = \frac{y}{\beta}$ to the range of $[0; 1]$ ($\beta = const.$).

The decay of the $1620\,nm$ overtone absorption is clearly seen in Figure 6.2, from an increasing radiant quotient $C_{acrylate}$. This decrease in absorption and increase in the radiant quotient $C_{acrylate}$ is directly connected to the degree of conversion, as mentioned before. The relative signal differences increase with the coating thickness from 0.7% to 6.3%. The given errors are determined by computing the standard deviation over 5 different samples at each thickness. In Figure 6.3 the same five samples were measured in the spectrophotometer to verify the acquired data from the sensor system. According to the above described method the discrete spectral intensities at $1620\,nm$ and $1550\,nm$ of the spectrophotometer were used to calculate the radiant quotient $C_{acrylate}$ of both polymerized and unpolymerized samples.

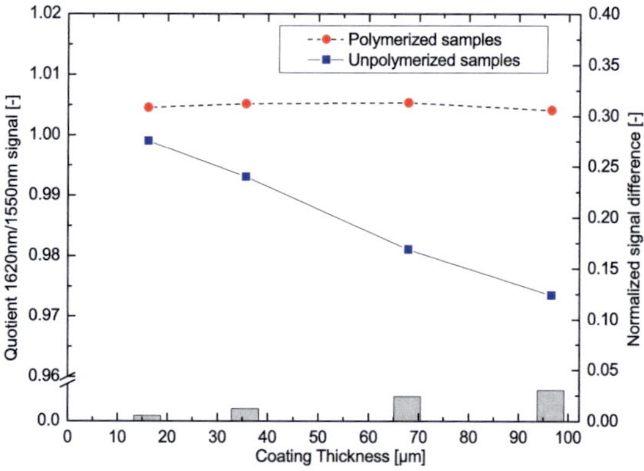

Figure 6.3: Measured signal results from unpolymerized and polymerized samples obtained by the spectrophotometer. Left ordinate shows the absolute radiant quotient of the sensor system. Right ordinate and bar chart shows the signal difference of the coatings before and after UV curing. The coating thickness of each measurement was determined after curing.

In Figure 6.2 the measured data of the discrete sensor system shows an unexpected maximum at the coating thickness of $67\,\mu m$. This maximum is a result of the scattering of the laser light in the coating layer. It is deflected from the incident light path. This radiant quotient value is saturated at a distinct point controlled by the layer thickness. Beyond this layer thickness the light propagation does not change any more and stable light scattering condition is reached. The reoccurring radiant quotient decrease at higher thickness values in the unpolymerized state is due to the occurrence of more acrylic $C-H$ compounds, thus resulting in an increased absorption at $1620\,nm$. The surface reflection of the system can be neglected in this case, due to the system inherent blocking of the direct reflected laser light. The data measured by the spectrophotometer in Figure 6.3 does not show this maximum in the radiant quotient. It was gathered by integrating over the whole half-space of the reflected light from the sample. This incidence is again proved by comparing the relative signal differences between polymerized and unpolymerized data of each measuring system. The signal difference range of the discrete sensor system is notably higher, due to the blocking of the direct reflected light. The result indicates one advantage of the laser sensor system concerning the possible conversion detection sensitivity with increasing coating thickness.

The measurement with the spectrophotometer is performed without motion of the sample. The low standard deviation of the spectrophotometer is due to the high precision of the device negligible and not displayed. The results of these verifying measurements are in great agreement with the results shown in Figure 6.2. The high standard deviation of the laser sensor systems on different samples is caused by the comparable rough steel substrates. Measurements performed on different uncoated substrates like paper or PVC indicate a lower standard deviation by one magnitude. Furthermore an even more controlled coating application, as printing processes provide will expect less errors.

6.2 Potentials and limits of the sensor

6.2.1 Lower coating thickness limit

The transflection principle of the sensor can be described as an integration of the signal loss by the C-double bonds in the coating. The incident laser light is partially absorbed on its way through the coating to the substrate and back to the surface. Therefore, a minimum of coating thickness exists, which is necessary to get an absorption signal which can be discriminated by the sensor system.

Limitations are the reactivity of the used coating, more specific the amount of C-double bonds, the scattering behaviour of the substrate and the overall sensitivity of the sensor.

The conversion quotient can be described in analogy to the Lambert-Beer Law (see Section 6.3) and the scattering behaviour of an arbitrary particle according to Bohren *et al.*, 1983 [66]:

$$C_{acrylate} = \frac{I^*_{1620}}{I'_{1550}} = \frac{I^* e^{-\alpha^* x}}{I' e^{-\alpha' x}} \Rightarrow C_{acrylate} = \frac{I^*}{I'} \cdot e^{-\alpha_c x} \qquad (6.1)$$

$$\text{with} \qquad \alpha^* = \alpha_c + \alpha_S + \alpha_p \quad \text{and} \quad \alpha' = \alpha_S + \alpha_p$$

with the coating thickness $x = 2 \cdot d$, polymer absorption coefficient α_p, a so called scattering absorption coefficient α_S and acrylic C-double bond absorption coefficient α_c. The index values in I^* and I' denote the laser wavelength in question.

The above stated analogy to the Lambert-Beer law is only valid with the assumption of single scattering and will be proved in the consecutive experiments. The conversion quotient can be calculated without knowing the explicit values of α^* and α'. The quotient is finally independent of the scattering and general polymer absorption. Only the characteristic absorption coefficient of the acrylic C-double bond at $1620\,nm$ is measured within this quotient.

The following extreme value consideration is useful for a better understanding:

- For an unpolymerized transparent formulation with infinite thickness:
  $\alpha_c \gg \alpha_p$ and $\alpha_c \gg \alpha_S$; $x \to \infty$
  $\Rightarrow C_{acrylate} = 0$

- For a polymerized transparent formulation (100% conversion) with infinite thickness:
  $\alpha_c = 0, \alpha_p > 0$ and $\alpha_S > 0$; $x \to \infty$
  $\Rightarrow C_{acrylate} = 1$

- For a polymerized and unpolymerized formulation with zero thickness:
  $x \to 0, e^{-\alpha_c x} = 1$
  $\Rightarrow C_{acrylate} = 1$

The minimum coating thickness can be estimated by a linear extrapolation of the polymerized and unpolymerized data, with the condition that $\alpha_c \cdot x \ll 1$:

$$C_{diff} = \left[1 - \frac{C_{acrylate}\,(unpolymerized)}{C_{acrylate}\,(polymerized)} \right] \tag{6.2}$$

with C_{diff} as the signal quotient difference with respect to above introduced assumption for 100% conversion. The extreme values for C_{diff} in case of two formulations a and b with $C^a_{acrylate} < C^b_{acrylate}$ are derived as follows:

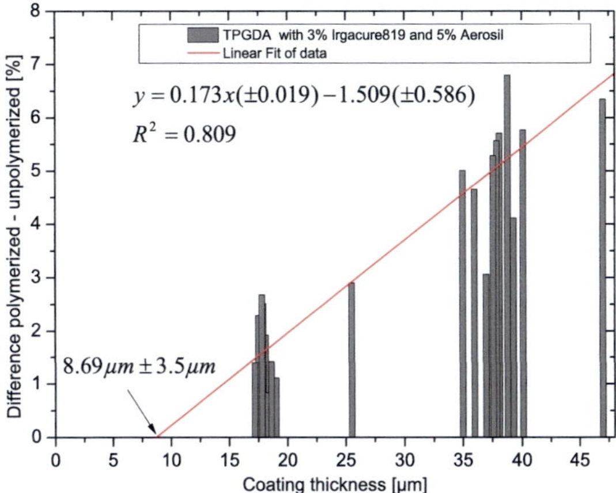

Figure 6.4: Calculation of the minimum coating thickness of TPGDA mixed with 5% Aerosil, which increases the viscosity for the application on the Q-panel substrate. The relative differences between $C_{acrylate}(polymerized)$ and $C_{acrylate}(unpolymerized)$ show the expected linear dependence on the coating thickness.

$$\text{For a coating thickness } x \to \infty : \quad C_{diff} = \left[1 - \frac{C^a_{acrylate}}{C^b_{acrylate}} \right] > 0 \quad (6.3)$$

$$\text{For a coating thickness } x \to 0 : \quad C_{diff} = \left[1 - \frac{C^a_{acrylate}}{C^b_{acrylate}} \right] \to 0 \quad (6.4)$$

With values of C_{diff} small enough a linear extrapolation is possible.

Figure 6.4 shows numerous measurements of the signal quotient differences between unpolymerized and polymerized samples of TPGDA with Aerosil at coating thicknesses of about $20\,\mu m$ and $40\,\mu m$. All coating thicknesses have been measured individually after the polymerization. The data of these measurements allows the determination of the minimum coating thickness by a linear extrapolation. The minimum coating thickness for the detection of a difference between an unpolymerized and polymerized sample is determined to $8.7\,\mu m \pm 3.5\,\mu m$. It is important to mention, that the minimum thickness result is only valid for the specific absorption coefficient α_c of the here analysed formulation.

6.2.2 Sensor performance: Long term stability and accuracy

As shown in Figure 6.5, an uncoated rough grey cast iron sample was measured several times to verify the signal stability and reproducibility. These measurements show a standard deviation of only 0.23% with all eight measurements. Thus the comparable rough surface of this sample hardly affects the sensor accuracy. The sensor system is independent of the surface roughness, by means of the normalization with the additionally applied reference laser. The computed standard deviation of the signal on the grey cast iron allows assuming an even higher accuracy of the sensor system on smooth substrates like any kind of paper or plastics.

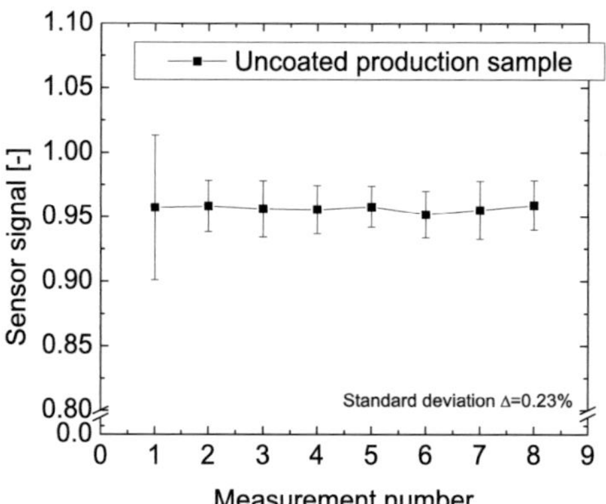

Figure 6.5: Sensor signal comparison of eight independent measurements on an uncoated grey cast iron production sample. The measurement deviation is calculated from the mean to $\Delta = 0.23\%$.

The sensor system must compensate the substrate surface roughness as well as the coating thickness variation. This thickness variation is highly dependent on the applied coating technique. The samples in Figure 6.6 were spray-coated and exhibit coating thicknesses between $28\,\mu m$ and $37\,\mu m$. Thus the standard deviation of the thickness is about $\Delta = 9.9\%$. An accurate discrimination between the unpolymerized and polymerized samples with the above stated signal differences of 3.5% to 5.5% would not be possible without correcting these deviations. As shown in Figure 6.6 the sensor system is capable of compensating these deviations similar to the discussed surface deviations. The resulting signal with a deviation of $\Delta = 1.9\%$ allows for the discrimination of the sample state again.

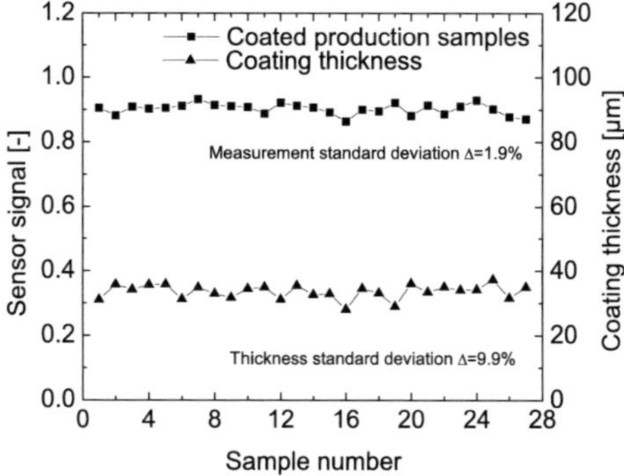

Figure 6.6: Measurements of 28 coated grey cast iron samples proofing that the coating thickness has no effect on the sensor signal. The signal deviation ($\Delta = 1.9\%$) of the sensor is about one order of magnitude less than the coating thickness deviation ($\Delta = 9.9\%$).

The performance of the sensor, regarding long term stability, is most important for the later purpose of a continuous monitoring in an industrial coating process. The weekly repeated measurements shown in Figure 6.7 reveal a mean deviation of approximately 2% within eight weeks. This deviation is compared to the later discussed change in the absolute signal far too high (see Section 6.3 and 6.4). The reason for this high value is identified in the Lock-In technique. The sensor system is switched off and on between the measurements, thus the Lock-In will always lock-in to a different phase part of the modulation, when it is turned on. The later developed prototype by QIAGEN comprises a digital Lock-In and omits this disadvantage by a phase monitoring of the modulation.

6.3 Signal reflection in the coating, penetration depth of the sensor

The main advantage of using lasers as light sources, in comparison with halogen light sources, is their high beam intensities. This enhances the depth of penetration into an acrylate coating. For this, gathering integral information of the polymerization conversion from the substrate coating boundary up to the surface is possible (see Pieke, 2010 [15]).

Four different acrylate formulations were analysed by using the depth profiling tool described in Section 5.5.1. The radiant quotient of the two laser

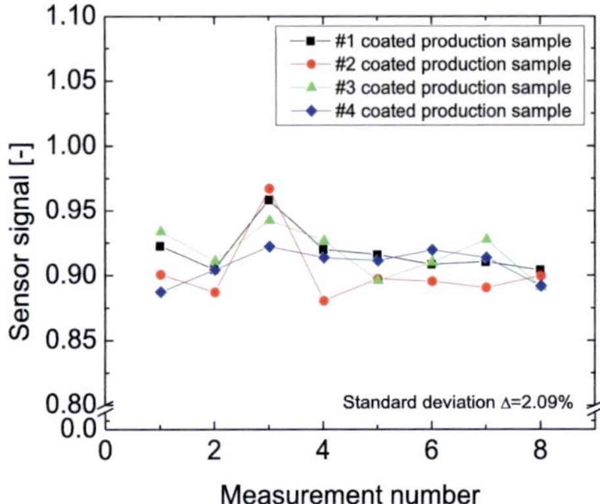

Figure 6.7: Long term measurement of four fully cured coated production samples. All samples were measured every week under the same conditions. The sensor shows a mean deviation of about 2%, over a period of eight weeks.

beam signals is plotted as function of the coating thickness of clear and pigmented formulations (Figure 6.8). All radiant quotients show an exponential decay with increasing coating thickness, according to the Lambert-Beer law [67]. By comparing the signal strength at $0\,\mu m$ of the clear formulation according to Figure 6.9 with the signal strength of Figure 6.2, it is ensured that the remaining coating thickness on the small boom substrate (see Figure 5.11) is less than $20\,\mu m$. The TiO_2 pigmented formulations show three different decay slopes as well as different minimum signals.

The pigmented formulations data were normalized with $y' = \frac{y - y_{min}}{y_{max} - y_{min}}$ for a detailed comparison and determination of the actual penetration depth. The normalization into the range of $[0; 1]$ is computed for a better comparison of all pigmented formulations with the assumption earlier discussed in Section 6.2.1 with the boundary conditions on page 70 in Equation (6.1). The normalized data are shown in Figure 6.9.

The Lambert-Beer law, as already mentioned in Subsection 6.2.1 is as

$$I = I_0 e^{-\alpha x} \tag{6.5}$$

with I for the measured light intensity, I_0 for the incident light intensity, α for the overall absorption coefficient and x for the light path distance in the material. In analogy to this law, all data were fitted using an exponential decay function

$$I = I_{Off} + Ae^{-\alpha x}. \tag{6.6}$$

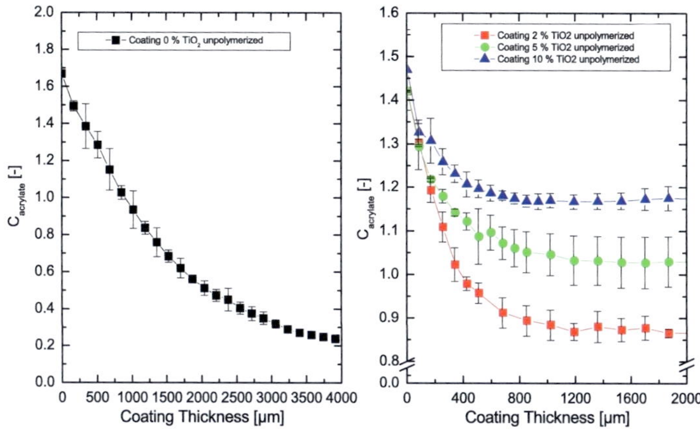

Figure 6.8: Penetration depth of the laser signal in clear and pigmented formulations. Left figure shows the clear unpolymerized formulation, right side shows the TiO_2 pigmented unpolymerized formulations with respect to the coating thickness. All curves show an exponential decay according to the Lambert-Beer law. Acquired signals by using the experimental set-up described above.

Where I_{Off} represents the intensity offset caused by the unwanted direct reflected light, A the related fitting coefficient (I_0 in the Lambert-Beer law) and I the measured laser light intensity weakened by the $1620\,nm$ absorption. The results are summarized in Table 6.1.

Formulation	10% TiO_2	5% TiO_2	2% TiO_2
RMS error	0.019	0.010	0.004
I_{Off}	0.014	0.039	0.069
Std. deviation	0.010	0.009	0.006
A	0.945	0.922	0.956
Std. deviation	0.031	0.022	0.015
$\alpha[1/cm]$	49, 4	35, 3	35, 3
Std. deviation	±3.258	±1.721	±1.167

Table 6.1: Exponential least square fitting results of the TiO_2 pigmented unpolymerized formulations using $I = I_{Off} + Ae^{-\alpha x}$ according to the Lambert-Beer law.

All RMS-errors of the least-square fit are below 0.02, which states a good model fit. I_{Off}, the undesired direct light reflection, is increasing from

0.014 to 0.069, due to the decreased scattering by less TiO_2 particles. The coefficient A with values of 0.92 to 0.95 indicates some uncertainties in the measurement owed to some minor remain of coating on the substrate at a thickness value of $0\,\mu m$.

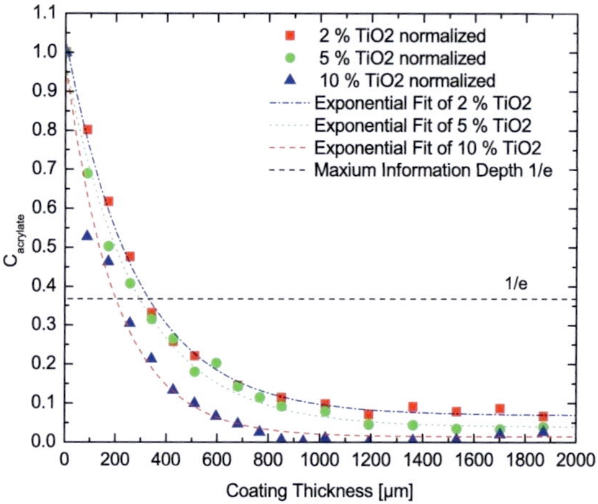

Figure 6.9: Normalized data of the measured unpolymerized pigmented formulations including the exponential fit of each dataset. The information depth at $1/e$ is plotted as well.

The calculated absorption coefficient α is, due to the nature of the used laser radiant quotient, the difference of the absorption coefficients $\alpha*$ and α' at $1620\,nm$ and $1550\,nm$, respectively.

The absorption coefficients are connected, as already discussed, according to Equation (6.1):

$$\alpha^* = \alpha_c + \alpha_S + \alpha_p \quad \text{and} \quad \alpha' = \alpha_S + \alpha_p \qquad (6.7)$$

The portion of each single coefficient can not be derived in this experimental set-up with respect to the C-double bond conversion and thus the variable $1620\,nm$ absorption coefficient α_c will not be further evaluated.

However the acquired absorption coefficients are in the same order of magnitude. Thus, these results are suited for calculating the more important penetration depth of the laser signals.

The determination of the maximum information depth X_{ID} is arbitrarily defined by Harrick and du Pre, 1966 [36] using the Lambert-Beer law and

Equations 3.18, 3.19 and 3.20 as follows:

$$I = I_0 e^{-\alpha d_p} \qquad \text{with} \qquad d_p = \frac{1}{\alpha} \cdots \rightarrow \frac{I}{I_0} = \frac{1}{e} \tag{6.8}$$

This definition is based on the exponential decay of the evanescent field strength with the depth d_p in the internal reflection spectroscopy. After the field strength hast dropped below e^{-1} no more significant information about the material can be collected. This argumentation is adapted to the here described experiment for the determination of the maximum information depth of the sensor.

The actual penetration depth of the laser system is determined from Figure 6.8. The maximum penetration depth is reached, as soon as there is no signal change with depth any more and the curve is getting constant. Approximately at $800\,\mu m$, $1200\,\mu m$ and $1400\,\mu m$ for the 10%, 5% and 2% formulations, respectively.

In contrast, the information depth d_p of the sensor system in 2, 5 and 10% TiO_2 pigmented coatings is calculated to $330\,\mu m$, $300\,\mu m$ and $200\,\mu m$ respectively. This gathered information of the coating layer is an integral over all infinitesimal thin layers between the surface and the substrate respectively the maximum penetration point. Any C-double bond or rather $C-H$ ligand interacting with the laser light will be accounted to the signal no matter of depth position.

The higher orders of the exponential fit, which would account for a better fit, and cause the offset I_{Off}, are neglected. If a reciprocal correlation between the pigment concentration and the penetration depth is assumed, we can extrapolate the data to higher concentrations. The extrapolated result leads to a maximum of $20\,wt\%$ concentration of TiO_2 at nearly zero microns.

The analysed samples with 2% to 10% TiO_2 pigments added, represent a common range of pigment proportions used in industrial applications, for example anti-corrosion coatings. Especially in this application area with complex geometries, varying irradiance power on the surface and different coating thicknesses, the information on remaining C-double bonds, which is a measure of non-conversion, is most desirable.

6.4 Conversion vs. irradiation time

The in situ measurement of the conversion within a $80\,\mu m$ thick coating layer is analysed by three independent methods: ATR-FTIR measurements, reflection of the coating using the photospectrometer and the transflection of the coating with the developed sensor system. A similar, more simple approach was conducted by Nakano *et al.*, 1993 [68], yet in the far infrared region. Furthermore, Decker *et al.* contributed with profound work to the understanding of the time dependency of the photopolymerization [64, 69, 47, 70, 11].

Fifteen coating samples were irradiated in time steps starting from $0.1\,sec$ to $1\,sec$ ($100\,msec$ time increments) and with $2\,sec$ time increments up to $10\,sec$. Each sample was investigated by all three methods. The acquired data are displayed in Figure 6.10, according to similar experiments found in literature e.g. Mijovic *et al.*, 19980 [71], Souza *et al.*, 2009 [72] or Decker *et al.*, 1992 [47]. The ATR-FTIR data are based on the C−H oscillation at $809\,cm^{-1}$ and $1407\,cm^{-1}$ and directly displayed in the Figure. The reflection data of the spectrophotometer are treated analogue to the data of the transflection sensor (see Equation 5.5). The signal behaviour in Figure 6.10 of all three methods is comparable. The amount of C-double bonds in the formulation is considerably decreasing with increasing irradiation time.

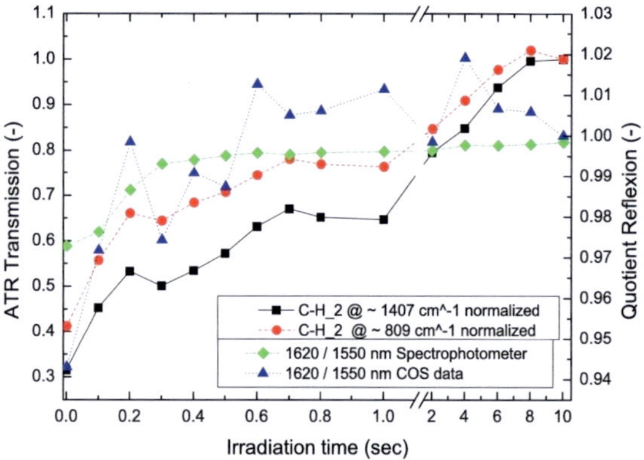

Figure 6.10: Overview of the acquired data during the irradiation time experiment of the three different measurement methods: ATR-FTIR absorption signals at $809\,cm^{-1}$ and $1407\,cm^{-1}$ and the photospectrometer and discrete laser sensor signals calculated according to Equation 5.5. The ATR-signals are displayed on the left ordinate. The $C_{acrylate}$ quotient of the discrete laser sensor and the photospectrometer are displayed on the right ordinate.

The reference degree of conversion $C[\%]$ of the acrylate system is calculated according to equation 5.7 with the spectrally integrated absorption values of the ATR-FTIR measurements at $810\,cm^{-1}$, $1407\,cm^{-1}$ and $1635\,cm^{-1}$. The integrals were computed for each absorption peak in the range of $790\,cm^{-1} - 819\,cm^{-1}$, $1395\,cm^{-1} - 1437\,cm^{-1}$ and $1626\,cm^{-1} - 1648\,cm^{-1}$, respectively. The acquired data from the discrete laser sensor and the reference data were adapted to the FTIR data. This adaption is necessary because of the non-vanishing character of the used overtone oscillation at $1620\,nm$. The resulting degree of conversion is displayed in Figure 6.11. The relative standard deviation of the discrete laser sensor is calculated from the data shown in Figure 6.12. The high standard deviation in the beginning of the irradiation experiment is caused by the minimum signal difference between the polymerized and unpolymerized samples. With increasing irradiation time the discrimination between the two sample states is increasing, thus the standard deviation is decreasing (see Figure 6.12).

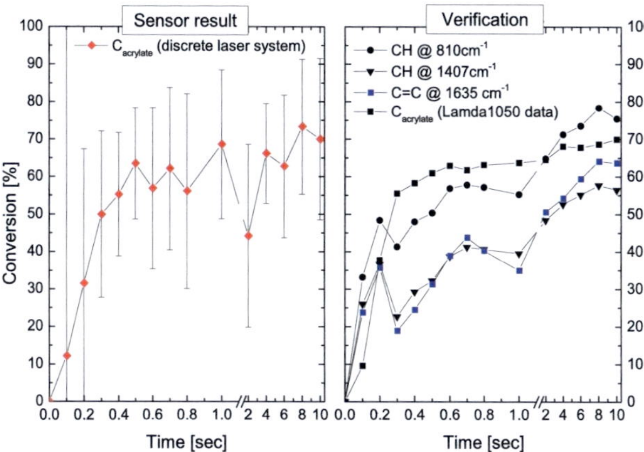

Figure 6.11: Comparison of time dependent acrylate conversion acquired with FTIR, spectrophotometer and discrete laser sensor data. The left figure shows the degree of conversion measured with the discrete laser sensor. The right Figure shows the corresponding measurements with the references, ATR-FTIR and spectrophotometer method. All data show a similar time-dependent raise of the conversion. The values $C_{acrylate}$ (discrete laser system) and $C_{acrylate}$ (Lambda 1050 data) were adapted to the range of the FTIR data.

The calculated degree of conversion of about $C[\%] = 70$ after 10 seconds is comparable with experimental values from literature [70, 14], whereas the degree of conversion in industrial applications is in the range well above $C[\%] = 80$ ([44, 15]. This instance is caused by the low irradiance power and the weak match of the photoinitiator absorption to the spectrum of the UV radiation source, respectively. Furthermore the discrepancy of the conversion values from the FTIR measurements at $810\,cm^{-1}$, $1407\,cm^{-1}$ and $1635\,cm^{-1}$ puts the assumption of complete vanishing of these absorptions in question. A sufficient answer to this discrepancy is not yet provided by literature. After an irradiation time of 300 milliseconds the two sample states can be discriminated.

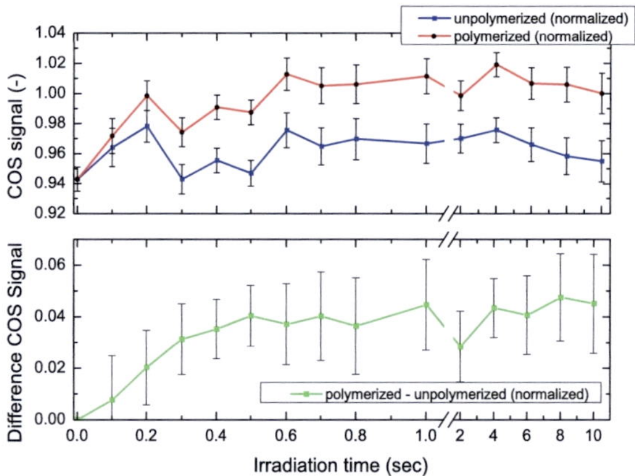

Figure 6.12: Experimental results with different irradiation times of the discrete laser sensor. Top figure shows the $C_{acrylate}$ quotient for the unpolymerized and polymerized samples. The bottom Figure shows the normalized difference between the two signals of the top figure, including the error propagation of each standard deviation. A discrimination of the unpolymerized and polymerized samples is possible after the first 300 milliseconds.

A detailed analysis of the results provided by the discrete laser sensor is shown in Figure 6.12. Both unpolymerized and polymerized signal quotients are displayed including the standard deviation calculated from the measurement track along $50\,mm$. The normalized difference of a unpolymerized and polymerized $80\,\mu m$ thick coating after 10 seconds is measured to $\approx 4.5\%$. This amplitude is used for the calculation of the degree of polymerization in Figure 6.11. The rather low resolution of the degree of conversion is caused by this small value in combination with the still high standard deviation. This result can be extrapolated to a lower coating thickness than $80\,\mu m$. As shown before in Section 6.2.1 we can assume a linear decrease of the absolute signal amplitude in with respect to the coating thickness. Thus, the expected absolute signal amplitude for a coating thickness of $80\,\mu m \cdot 0.2 = 16\,\mu m$ is $4.5\% \cdot 0.2 = 0.9\%$ with a degree of conversion of 70%. Always keeping in mind, that this result is only valid for the used coating formulation described in Section 4.1. Higher signal amplitudes are acquired with pure TPGDA mixed with Aerosil (see Figure 6.4).

6.5 Reflection and scattering on acrylate coatings with substrates

The scattering behaviour of coated and uncoated substrates like paper, metal or cardboard is essential for the performance of the spatial filter used in the sensor system (see Section 5.2.2). Depending on the scattering behaviour of the substrate and coating, the amount of radiation, which is guided onto the detector changes [73]. If enough laser power is available, this behaviour is less important but should be taken into account for determining the optimal distance between the sensor head and the sample. The radiant intensity measurements are performed with a $850\,nm$ Fabry-Pérot laser diode and a simple IR-enhanced VIS photo diode. The samples were illuminated by the laser diode and their transflected radiation detected under angles of $-4°$ to $14°$ measured against the vertical axis to the sample. The transflected or reflected beam profile is expected to be a combination of a cosine Lambertian function and a Gauss profile. Figure 6.13 shows the cosine reflection of an uncoated commonly used $80\,g/m^2$ white copy paper. This measurement also demonstrates the accuracy of the built measurement system.

The fit is calculated by using a simple cosine function

$$B \cdot cos\left(\frac{\Phi}{\beta}\right) \qquad (6.9)$$

with B as global Amplitude, Φ as observation angle and β as a correction factor for deviation of the pure cosine . The fitting result is summarized in Table 6.2. As expected, the reflection of paper is close to the assumed Lambert radiator.

A different behaviour of the reflection is observed at steel panels used in the experiment. The reflection property of steel is known to be highly dependent on the surface finish, like polishing, machining or roughening. Two different steel panels from Q-Labs© were measured in this set-up:

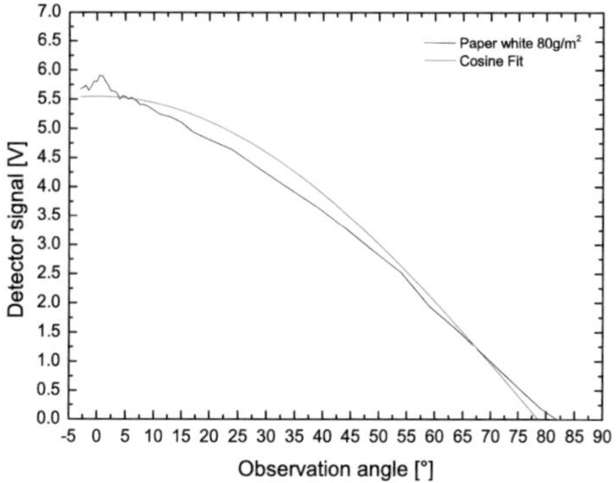

Figure 6.13: Acquired scattering data of a standard white copy paper. The Cosine fit (Equ.6.9 is plotted in red. A Lambert radiation characteristic is clearly observed with a corr. R^2 of 0.986 of the fit.

FP	White paper $80\,g/m^2$	
	Value	Std. dev.
B	5.548	0.0361
β	1.146	0.0112
corr. R^2	0.986	

Table 6.2: Summarized result of the fit displayed in Figure 6.13.

Type QD (smooth finish) and Type R (dull finish). The results are displayed in Figure 6.14. Both types show a Gaussian profile, whereas the reflection profile of the rough panel (Type R) is distinctively wider than that of a smooth panel.

All fitting results in Figure 6.14 and 6.15 are based on the assumption of combined scattering behaviour with parts of Lambert and Gaussian scattering. Equation 6.10 is the result of a deterministic fitting approach gaining the best overall result for the samples. The acquired values from Figure 6.14 are summarized in Table 6.3. The fit is designed to differentiate between the Gauss and Lambert scattering; the parameters A and B are a direct measure of each contribution, respectively. The Cosine share from the rough steel panel (Type R) is about $\approx 17\%$ whereas it is only about $\approx 6\%$ from the smooth panel (Type QD).

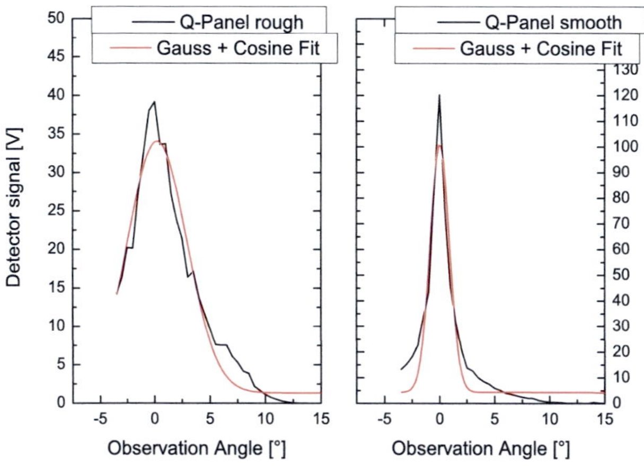

Figure 6.14: Reflection data of two different Q-Panels, Type R (rough) on the left and Type QD (smooth) on the right. The red line displays the fitting result according to Equation 6.10. Details of the fit are summarized in Table 6.3. The rough steel panel exhibits a broad Gaussian shape; the smooth panel shows a three times higher absolute amplitude and only about half of the scattering angle width.

FP	Steel panel rough		Steel panel smooth	
	Value	Std. dev.	Value	Std. dev.
A	7.75629	0.48285	69.78485	6.50981
σ	2.68298	0.11802	0.88255	0.0517
μ	0.17892	0.09658	-0.01903	0.04961
B	1.35183	0.52366	4.28745	1.19366
corr. R^2	0.96344		0.92782	

Table 6.3: Summarized result of the fit displayed in Figure 6.14.

$$y = A \left[\sigma \cdot \underbrace{\frac{1}{\sqrt{2\pi}} e^{-\frac{1}{2}\left(\frac{\Phi-\mu}{\sigma}\right)^2}} \right] + \underbrace{B \cdot cos\left(\frac{\Phi}{180/\pi}\right)} \qquad (6.10)$$

$$\underbrace{\phantom{A \left[\sigma \cdot \frac{1}{\sqrt{2\pi}} e^{-\frac{1}{2}\left(\frac{\Phi-\mu}{\sigma}\right)^2} \right]}}_{Gauss\ part} \quad \underbrace{\phantom{B \cdot cos\left(\frac{\Phi}{180/\pi}\right)}}_{Cosine\ part}$$

Figure 6.15 shows the reflectance of four Q-panels (Type R) coated with different industrially used unpolymerized acrylate coatings with typical coating thicknesses. The scattering data and its corresponding fitting results (Table 6.4) reveal almost exclusive Gaussian scattering behaviour. The ratio of the parameter B for Cosine behaviour to the corresponding parameter A (Gaussian part) is in three of the four cases only about $\approx 1\%$. Only the transparent acrylate overprint shows a ratio of about $\approx 6\%$. Due to the overprint coating thickness of $20\,\mu m$ the scattering effect of the rough steel panel is still measureable in this case. The blue overprint with the same coating thickness is already covering this effect. The only difference between those two formulations is the addition of blue pigments.

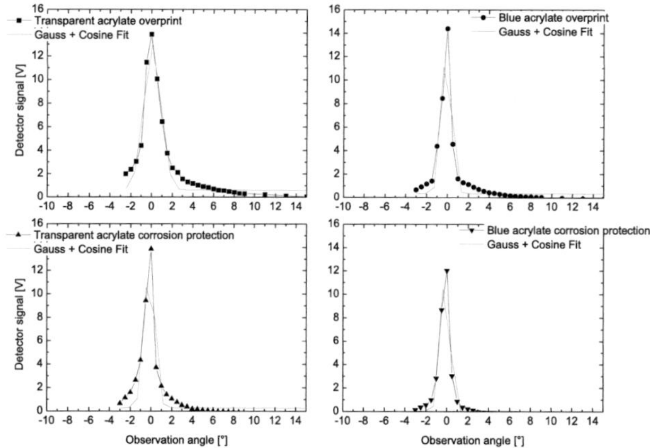

Figure 6.15: Scattering of four different acrylate based coatings used in the automotive and printing industry applied on Q-panels (type R). All four samples show a Gaussian scattering profile. The fitting results are summarized in Table 6.4. The broad Gaussian peak from the uncoated rough steel panel is sharpened as coating is applied.

For increased coating thickness of formulations for corrosion protection up to about $60\mu m$, the same effect is revealed. The small Cosine to Gaussian portion of 1.5% in the transparent coating, vanishes with the addition of blue pigments completely.

The scattering analysis of the different substrates and coating formulation shows a trend to more Gaussian scattering for higher coating thicknesses and coloured coatings by application of pigments. Pure polymer formulation have been studied with neutron scattering and showed a similar behaviour with increasing molecule chain length (see Fixman *et al.*, 2004 [74]).

The angular scattering result at $\pm 2°$ can be used for the calculation of the radius of the Gaussian scattering cone ($r = d \cdot sin(2°) = 1.75\,mm$ at a distance of 50 mm. This distance was experimentally found to be a good

FP	Transp. acrylat overprint		Blue acrylate overprint	
	Value	Std. dev.	Value	Std. dev.
A	9.60172	0.7772	18.56514	1.61131
σ	0.82007	0.04203	0.4586	0.02456
μ	0.03306	0.03997	-0.12239	0.02405
B	0.61507	0.14624	0.34453	0.11639
corr. R^2	0.94863		0.94238	

FP	Transp. acrylate corr. protection		Blue acrylate corr. protection	
	Value	Std. dev.	Value	Std. dev.
A	16.81763	1.70045	18.2624	0.95454
σ	0.49134	0.03075	0.43267	0.01397
μ	-0.17874	0.02979	-0.17213	0.01324
B	0.26835	0.13621	-0.00333	0.06126
corr. R^2	0.92495		0.97981	

Table 6.4: Summarized result of the fit displayed in Figure 6.15.

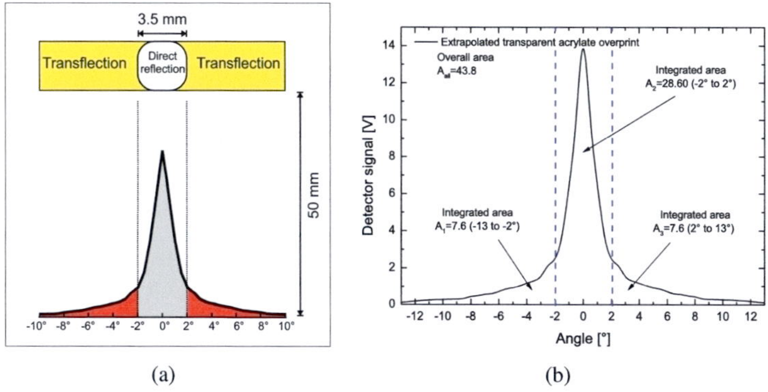

(a)   (b)

Figure 6.16: Scattered radiation ratio used for the detection of the conversion. (a) Principle set-up of the spatial filter in relation to the scattering profile of a transparent acrylate formulation on a Q-panel (Type-R).(b) Approximated ratio of the radiation used for the detection and the ratio of the loss trough the mirror hole.

compromise for the signal quality of all sample measurements. The main direct reflection of the laser beam on the coating surface interacts only in the first few micrometers with the coating. The other transmitted part of the main direct reflection will be reflected from the substrate in the same direction and cannot be isolated from the first part of the surface reflection. Thus the direct reflection does not contain any important conversion information and is not useful for the conversion detection.

Figure 6.16 shows the principle of the scattered radiation collected by the spatial filter and an approximation of the received radiation ratio of the overall signal. The main peak integrated area is 28.6. Whereas the slope parts, received by the detector is calculated to $2 \cdot 7.6 = 15.2$. Only less than $\approx 35\%$ of the overall radiation is used for the detection.

Chapter 7

Prototype development

The laboratory discrete laser sensor has been transferred to a smaller unit suitable for the use in industrial environments. This transfer has been realized by QIAGEN Lake Constance GmbH within the BMBF project. During this process further tests with coating formulations were conducted. The gained results gave reason for the extension of the reference measurement principle by adding a third laser source. Section 7.1 explains the reasons and introduces the described reference extension. Section 7.2 presents the final prototype, as used in the industrial environment for the field testing. The mechanical part as well as the software are explained.

7.1 Improvement of the detection reference due to additional experimental results

The earlier used two laser detector principle has been implemented into the first prototype and thoroughly tested within simulated industrial conditions, including industrially used formulations and coating thicknesses. A high power UV-curing unit was used in combination with a linear moving stage for velocities up to $300\,m/min$ at the IST Metz GmbH laboratories. During this testing the conversion detection of thick coatings in the range of several ten microns was proved. However, the detection failed for some industrial coatings. Main differences between the model formulation and the industry formulations are the additives and pigments as well as a less reactivity due to highly prepolymerized acrylates. These differences cause an unexpected spectral behaviour of the formulations. The boundary condition of a constant reference at $1550\,nm$ is failing. The spectral transmission shows a minor global shift in the slope characteristic between the unpolymerized and polymerized state. Figure 7.1 shows the transmission spectra of a gloss coating before and after being UV irradiated. A linear fit between $1400\,nm$ and $1600\,nm$ is added to explain the spectral slope change, causing the failing of only one reference for the detection of polymerization. The slopes of both fits differ by 15% (absolute $0.0016 \pm 4.23e^{-5}$) related to the fit of the unpolymerized coating.

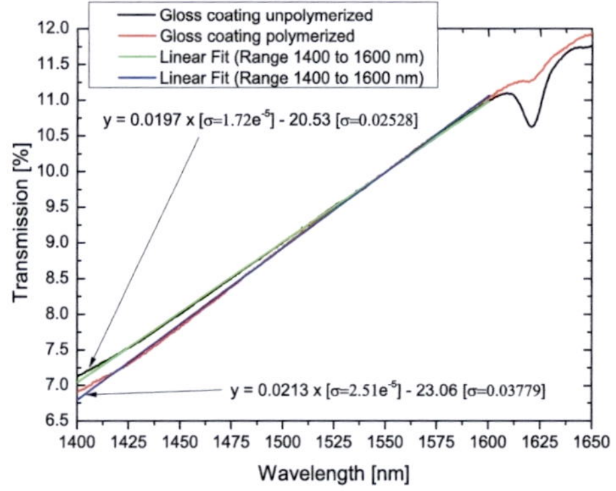

Figure 7.1: Observation of the spectral slope change of a gloss coating (thickness of more than $50\,\mu m$) before and after irradiation. The linear fit was calculated for the spectral data from $1400\,nm$ to $1600\,nm$. The linear equations for both states are displayed. The slope change is calculated to 15% (absolute $0.0016 \pm 4.23e^{-5}$) based on the unpolymerized fit.

The observed change by different slopes in Figure 7.1 is small, compared to the absorption change at $1620\,nm$. The values of the linear fit extrapolated to $x = 1620$ are $y_{unpolymerized}$=11.384% and $y_{polymerized}$=11.446% whereas the actual absorption value at $1620\,nm$ is 10.63%. In this case a distinct differentiation is still possible. Always keeping in mind that the conversion is measured on a coating thickness of more than $50\,\mu m$, while realistic thicknesses are in the range of $\approx 20\,\mu m$. Figure 7.2 shows measured spectra of two anti corrosion coatings (acc) and one transmission spectrum of TPGDA in unpolymerized and polymerized states, respectively. Furthermore the Figure includes a detailed view on the hardly visible $1620\,nm$ absorption of the anti-corrosion coatings.

The earlier discussed slope change during polymerization considerably influences the measurement results of the anti corrosion coating in Figure 7.2. Table 7.1 shows the linear fit of all four coatings in the range of $1400\,nm$

to $1600\,nm$. A slope increase by more than 150% and 250% for both types of coatings, black and grey, from 0.0145 to 0.0225 and 0.0057 to 0.0152 is determined, respectively. These spectroscopic measurements are used for the demonstration of the advantages of a second reference wavelength.

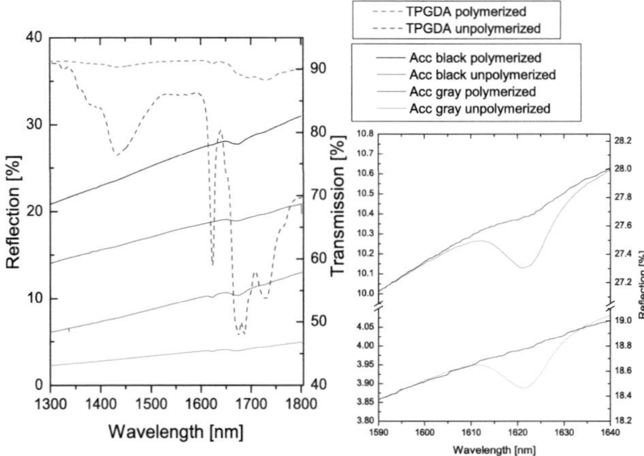

Figure 7.2: Reflection from industrial anti- corrosion protection coating (acc) ($24\,\mu m$ coating thickness) before and after irradiation in the left graph, left ordinate. In comparison from unpolymerized and polymerized transparent TPGDA the transmission is shown on the right ordinate, left graph. For better visibility a detailed view on the $1620\,nm$ absorption band of the anti-corrosion coatings is shown in the right graph. The signal change of the absorption is determined to $\approx 0.2\%$ and $\approx 0.1\%$ for the black and grey anti corrosion coating, respectively. The slope change is slightly visible. For a clear comparison Table 7.1 comprises all linear fitting results of the four anti corrosion coatings.

Figure 7.3 schematically shows the measurement principle implemented by the use of a third wavelength. The virtual reference value 3 is derived from the calculated slope by means of reference 1 ($1530\,nm$) and reference 2 ($1570\,nm$) extrapolated to the absorption band at $1620\,nm$. The virtual reference is compared with the actual measurement value at $1620\,nm$ and its difference is calculated. This principle compensates not only for coating thickness changes or substrate deviations compared to the system with one reference. By introducing one more independent value another unknown effect, the discussed slope change, is corrected. The amount of additional work and expenses for the built prototype with three laser sources is rather low due to the flexible set-up. More details on the prototype set-up are described in Section 7.2.

Sample name	y-axis intercept	std. error	slope slope	std. error	corr. R^2
Acc black polymerized	-8.63	0.039	0.0225	2.65E-05	0.99
Acc black unpolymerized	-12.98	0.023	0.0145	1.56E-05	0.99
Acc grey polymerized	-5.72	0.040	0.0152	2.67E-05	0.99
Acc grey unpolymerized	-5.24	0.007	0.0057	4.69E-06	0.99

Table 7.1: Linear fit of the anti corrosion coatings (Figure 7.2) in the spectral range of $1400\,nm$ to $1600\,nm$. The slope change in both coating fits, grey and black, before and after polymerization is clearly visible. The slope of the black and grey coating increases more than 150% and 250%, respectively.

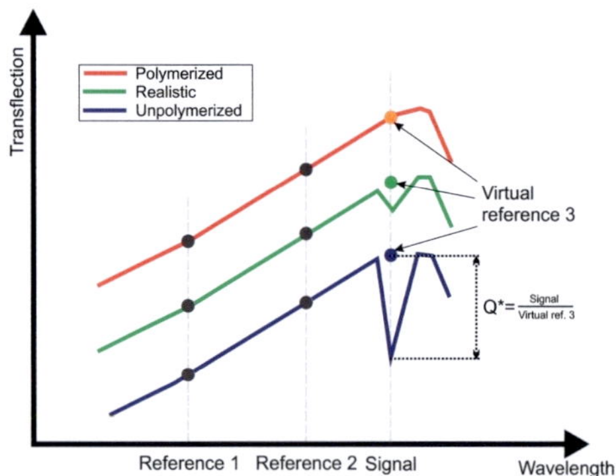

Figure 7.3: Measurement principle of the discrete laser sensor with two reference laser sources and the detection laser source. The resulting slope from reference point 1 and reference point 2 is used for computing the virtual reference 3 at the signal wavelength. The virtual reference is compared to the actual transflected signal and the quotient Q^* is acquired.

The effect of the third wavelength is calculated in Table 7.2 on base of spectroscopic data. In comparison Table 7.3 summarizes the results of measurements with only one reference. The implementation of the third wavelength leads to the following signal processing:

$$\text{Reference slope:} \qquad m_{\text{Ref}} = \left[\frac{\text{Signal}_{\text{Ref}_1} - \text{Signal}_{\text{Ref}_2}}{\lambda_{\text{Ref}_1} - \lambda_{\text{Ref}_2}} \right] \tag{7.1}$$

$$\text{Y-axis intercept:} \qquad y_{0_{\text{Ref}}} = \left[\frac{\text{Signal}_{\text{Ref}_1}}{m_{\text{Ref}} \cdot \lambda_{\text{Ref}_1}} \right] \tag{7.2}$$

$$\text{Linear extrapolation:} \qquad y^* = m_{\text{Ref}} \cdot \lambda_{1620nm} + y_{0_{\text{Ref}}} \tag{7.3}$$

$$\text{Virtual reference to signal quotient:} \qquad Q^*_{acrylate} = \frac{\text{Signal}_{1620nm}}{y^*} \tag{7.4}$$

The reference slope and the y-axis intercept are calculated by using the two reference wavelengths at $1530\,nm$ and $1570\,nm$ independently of the absorption wavelength at $1620\,nm$. The linear equation is extrapolated to $1620\,nm$ and a virtual reference is generated. With the current value of the absorption band and the virtual reference the $Q*$ quotient is calculated.

Above described equations 7.1 to 7.4 lead to a signal difference between the unpolymerized and polymerized black and grey anti corrosion samples of 2.48% and 2.72%. The former approach without the slope correction results in a difference of 2.17% and 1.97%. The grey formulation contains more than 40% of pigments and additives, whereas the black formulation only contains about one fifth of pigments and additives. This result reveals the necessity of the slope correction especially for formulations containing high amounts of pigments and additives. The absolute conversion is underestimated by the formerly used two wavelength approach. Furthermore the currently described three wavelength approach increases the accuracy and the stability of the measured conversion signal for formulations containing high amounts of pigments and additives.

Sample name	Reflection at $1530\,nm$	Reflection at $1570\,nm$	Reflection at $1620\,nm$	y-axis intercept	slope
Acc black poly.	25.84	26.71	27.64	-7.218	0.0216
Acc black unpoly.	9.14	9.72	10.14	-13.129	0.0146
Acc grey poly.	17.54	18.14	18.76	-5.123	0.0148
Acc grey unpoly.	3.52	3.75	3.89	-5.023	0.0056

Sample name	Extrapol. value at $1620\,nm$	Quotient $C^{*}_{acrylate}$	Quotient $C^{*n}_{acrylate}$	Difference	
Acc black poly.	27.79	0.9949	1	2.48%	
Acc black unpoly.	10.45	0.9703	0.9752		
Acc grey poly.	18.88	0.9938	1	2.72%	
Acc grey unpoly.	4.03	0.9668	0.9728		

Table 7.2: Results of the discrete laser sensor equations 7.1 to 7.4 with the spectrophotometric data displayed in Figure 7.1. With the compensation of the slope change a signal difference of 2.48% and 2.72% is calculated for the black and grey formulations, respectively. Table 7.3 summarizes the former quotient result with one reference.

Sample name	Reflection at $1550\,nm$	Reflection at $1620\,nm$	Quotient $C_{acrylate}$	Quotient $C^{n}_{acrylate}$	Difference
Acc black poly.	26.27	27.64	1.052	1.000	2.17%
Acc black unpoly.	9.43	10.13	1.074	1.021	
Acc grey poly.	17.86	18.76	1.050	1.000	1.97%
Acc grey unpoly.	3.63	3.89	1.070	1.019	

Table 7.3: Result of the formerly used two wavelength approach on the spectroscopic data. The polymerization difference of the grey formulation with high amounts of pigments and additives is corrupted by the slope change.

7.2 The discrete laser sensor prototype for industrial field tests

The sensor prototype as displayed in Figure 7.4 is fully operational and equipped with standard interfaces and IP 54 protected against dust and other pollutants. The system operates with 24 Volts, which is an industry typical supply voltage. The sensor head of the final three wavelength prototype only comprises rigid optics and mirrors and is connected via two optical fibres within a ruggedized shell cable. A detailed description of the sensor head has not yet been approved for publication by the manufacturer. The patent disclosure is pending.

For IP 54 protection, the main unit must not have any cooling fans and is passively cooled. The main unit is designed to fit in any 19 inch rack for the most possible flexibility. The requirements for laser security as demanded by the end user are satisfied by a machine interlock connection and customized metal shielding around the laser head. Without any additional protection the main unit plus sensor head is classified as a class 3B laser system. Most companies are requiring the lowest laser class 1 or 1M.

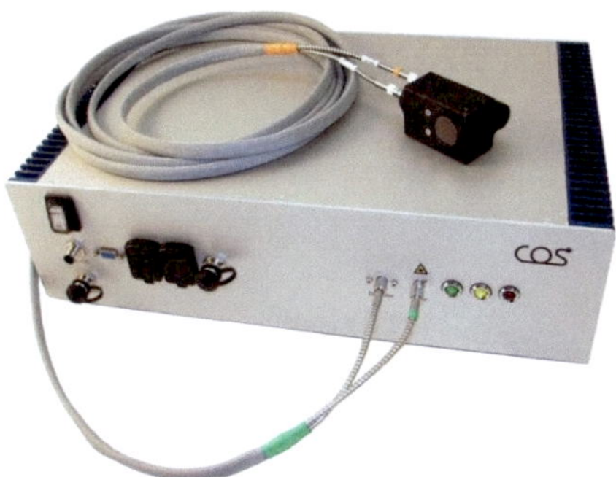

Figure 7.4: Final design of the sensor prototype developed within the BMBF project "CuringOnlineSensor". The sensor head is connected with optical fibres to the processing unit. Laser sources, InGaAs detector and power management are hosted inside this processing unit. The detector head only consists of fixed optical elements, thus suited for rough environments. (Photo courtesy of QIAGEN Lake Constance GmbH, Elmar Feuerbacher.)

The temperature and current control loop is implemented for each laser separately. The modulated laser light is combined by an optical element and is guided into the detector head. The transflected light from the substrate is collected and guided back to the controller. The resulting signal in the main unit is pre-amplified and converted for further processing. The Lock-In amplifier finally delivers three independent laser intensities for computation according to Equations 7.1 to 7.4. The necessary statistics for a final reduction of the standard error are adjusted for each application.

The acquired conversion data is visualized in a user friendly software interface, where all important information is summarized. A status light with red, orange and green indicates the current conversion of the product. Red light will lead to an immediate action by the operator or an automatic stop of all affected production processes. Orange indicates a serve drop in the conversion and thus in the product quality. An appropriate countermeasure should be initiated by the operator. As long as the conversion is above a pre-defined lower limit, the green status colour will flash up and the production is not interrupted.

7.2.1 Example measurement of the prototype

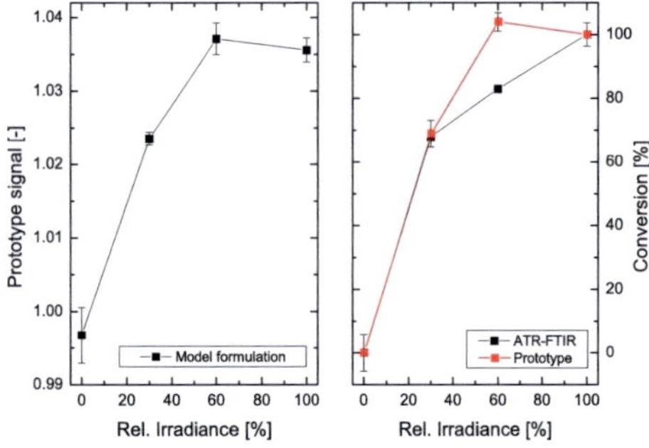

Figure 7.5: Measurement result of the prototype and the calculated conversion. The ATR-FTIR reference is plotted for comparison. The coating was applied with a thickness of $40\,\mu m$. The conversion of the prototype is in good agreement with the reference for 30% irradiance and is overestimated for a relative irradiation of 60%.

The prototype was tested in the laboratory in combination with an industrial medium pressure Hg-lamp and a conveyor belt speed of $100\,m/min$. The irradiance power of the lamp was varied (0%, 30%, 60% and 100%) and the conversion of each sample is measured five times with the discrete laser system at a speed of $40\,m/min$. For referencing, all samples were cross checked with a ATR-FTIR "Alpha" from Bruker [75]. The $1406\,cm^{-1}$

absorption peak was used to calculate the reference conversion in the far infrared (see Table 4.1). Figure 7.5 shows the result of the prototype measurement and the calculated conversion according to the prototype data and the ATR-FTIR reference. The absorption at $1406\,cm^{-1}$ is connected to the acrylate C-double bond (see Table 7.3) and therefor well suited for the conversion detection as this absorption is vanishing completely with a conversion value of 100%. Another possible absorption connected to the C-double bond of the acrylate is $810\,cm^{-1}$ (see publications [65, 46, 76] or [77]). The derived conversion of the prototype is in good agreement with the ATR-FTIR measurement. An almost identical result is achieved with an irradiance of 30 %, whereas the conversion is overestimated by the prototype at 60% irradiance. The standard deviation for the partially polymerized measurements is in the range of less than 0.0021 which corresponds to a conversion error of less than 6% for an absolute absorption difference of 0.0388 in the range of 0% to 100% irradiance. The highest deviation of $\pm$ 9.77% is measured with the unpolymerized sample .

Chapter 8

Prototype testing in industrial environments

The final prototype is installed for testing purposes in an industrial screen printing machine, used for the production of packaging. The packaging is conventionally printed with solvent based inks, only in a last printing step, an haptic UV topcoat is applied. This top coat is meant to increase the quality rating of the product. The formulations used, may differ in the amount of added pigments and thus in the degree of gloss. The printing is applied on an endless paper web, approximately 50 cm wide. Due to the geometry of the haptic printing and manufacturer conditions, the actual conversion measurement is realized on a test pattern at the margin of each package sample at a web speed of approximately 200 to $250\,m/min$.

All data are stacked, interpolated and a floating average conversion value is computed for a predefined amount of test patterns. The main goals of this prototype test are results for the overall performance and stability of the system under real production environments. The effects of high amounts of paper dust and solvent on the sensor system are examined and monitored over some month. As well as the accuracy of the trigger system, used for the identification of the test pattern data, is tested.

8.1 Installation of the prototype on a screen printing machine

The sensor system is located directly behind the UV-radiation unit, which is the last part of the printing process. Afterwards the finished printing product is checked for possible ink spilling and then stamped into single packaging pieces.

The sensor head as shown in Figure 8.1 is positioned normal to the paper web on an adjustment rig. The shielded optical fibres are visible in the foreground connected to the sensor head mounted on the adjustment plate. The complete system is movable across the width of the paper web to be adaptable to different layouts and test patterns. The sensor head and all the peripherals is shielded afterwards by a metal cover to fulfil the laser safety regulations.

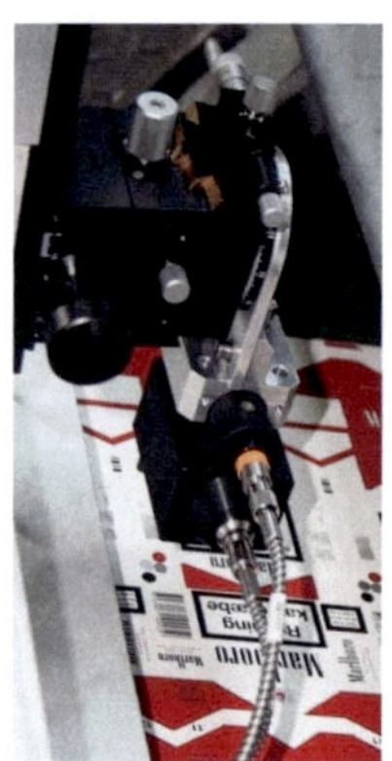

Figure 8.1: Mounted sensor head in the production. The sensor head is accurately adjusted perpendicular to the paper web by the use of the optical mount system. The two shielded optical fibres connected to head and processing unit are visible in the foreground.

Figure 8.2 shows the adjusted sensor head on the down turned paper web measuring the test pattern on the product.

The complete sensor system is connected to the printing machine and only operational if the UV-radiation and the UV-printing unit is operational as well. The conversion result is monitored and recorded on site in the processing unit for test purpose and later analysis. Yet, all necessary interfaces

Figure 8.2: Integrated sensor head on the printing machine with a metal cover (right side) and installed optical fibre connection.

and connectors are already available for the advanced system integration as it is required by the printing industry.

8.2   In-line measurement

The sensor measures continuously on the paper web to ensure a high stability of the laser sources, which would not be given by fast on and off switching. Therefore, with a web speed of more than 200 m/min and the small test pattern, a highly accurate trigger management is necessary for the selection of the actual test pattern data.

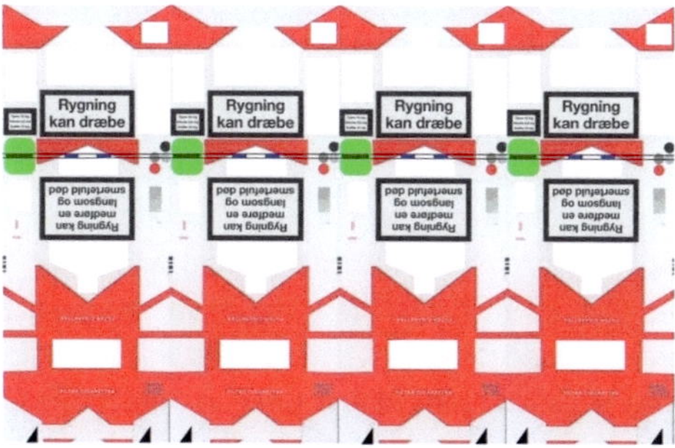

Figure 8.3: Example of the test measurement on the paper web. The laser line of the sensor is marked from left to right. The conversion measurement is performed on the green coloured test patterns. The blue areas are used for calibrating the sensor before each conversion measurement. The trigger marks of the printing machine, which are used for correct positioning of the sensor, are visible as black triangles at the bottom. A detailed view of the test pattern and the reference spots are shown in Figure 8.4.

The test pattern has a size of 11 x $10\,cm$ and its position is predefined by the manufacturer. The coating thickness on the test pattern is set to $\approx 10 - 20\,\mu m$. The comparable large range of the thickness value, is due to several production circumstances, e.g. wear out of the printing cylinders and the type of used UV coating formulation. With the paper web velocity of $250\,m/min$ each test pattern comprises 12 independent measurements, each over a length of $0.65\,mm$ ($\sum 7.80\,cm$). A margin of $1\,cm$ at the beginning and the end of the test field is used to buffer the data acquisition of the measurements.

First field tests gave reason for a reference measurement on the uncoated paper in close vicinity to the actual measurement pattern. The measured conversion values were not constant for time periods of more than one hour, due to unknown effects caused by the production environment. The sensor system itself was checked on this effect in the laboratory and did not show that variance under a controlled environment.

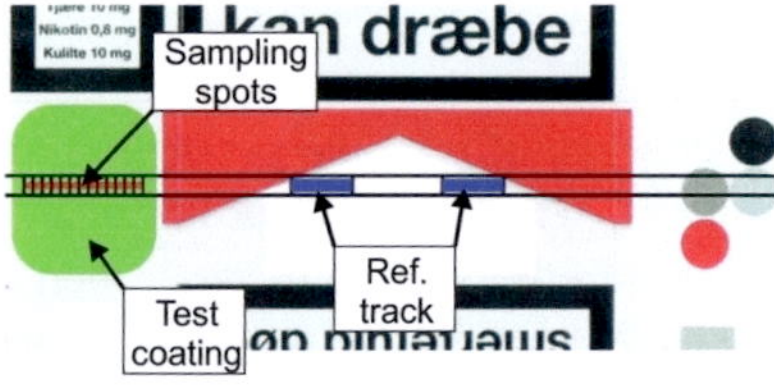

Figure 8.4: Detailed view of the conversion measurement on the paper web. The conversion measurement is realized on the green coloured test coating, which represents the actual haptic print on the packaging. All sample points on the test pattern, as well as the reference track are averaged and state for one measurement.

As a consequence of this result the mentioned reference field was introduced on a non printed stripe within the measurement track. By this method, the sensor is calibrated to a virtual 100% conversion value on this field, as there is no acrylate coating and thus no detectable C-double bonds, which could affect the measurement.

With the described set-up of the sensor (Section 8.1) and the above explained measurement program, the printing on the packaging was monitored for several hours, each time when a UV-coating print job was carried out. An example of the acquired data is shown in Figure 8.5. The sensor values and the standard deviation is plotted over a period of 29 hours. The mean value of the conversion quotient $C_{acrylate}$ is calculated to 0.955 ± 0.0052 and the variance to $2.667e - 5$.

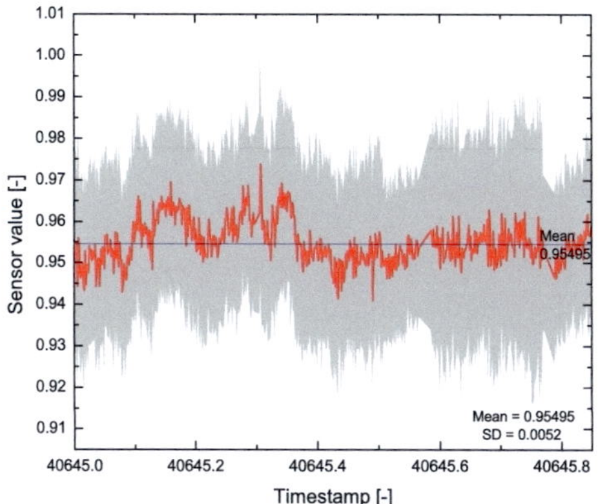

Figure 8.5: Field test data of the sensor system installed in an industrial screen printing machine. The conversion quotient $C_{acrylate}$ is plotted over a time of 29 hours (red line). The standard deviation of each measurement is plotted as a grey envelope. The average conversion quotient is calculated to 0.954. The standard deviation of the data in 29 hours is calculated to 0.0052.

This first field test within a realistic production environment disclosed several aspects regarding the sensor performance: During the field test duration common quality tests were conducted on the production results and none of the product was faulty. This result of the established quality control mechanisms is approved by the sensor signal. Still, the biggest drawback for the field test result is the lack of data from unpolymerized or at least partially unpolymerized production results. This lack is owed to the complex peripheral situation of the associated partners of the CuringOnlineSensor project. A production run with completely shut down UV-light sources can not be realized, still experiments with less inertial gas concentration and reduced lamp power should be possible for a detailed study of the sensor signal.

Chapter 9

Other advanced prospectives for the sensor system

The feasibility of unique monitoring of acrylate conversion for simple coating processes as screen printing has been shown in the previous chapters. Yet there is still development potential for various applications exceeding the current possibilities of the sensor system. First and foremost the combination of the conversion monitoring with already existing quality monitoring tools like image recognition or 3D scanners has to be mentioned for future developments of this measurement principle (Section 9.2).

Other application scenarios with different products are not yet evaluated and cross checked for economical benefits using this method. For example circuit board solder resist coating or UV-adhesive applications were identified as future high potential application areas for the sensor system during the development process.

The most important improvement in near future will be discussed in Subsection 9.2.1. A direct comparison between the unpolymerized and polymerized state of each measured sample.

9.1　Future product application areas

9.1.1　UV-glueing applications

Today, an increasing number of optical components used for consumer products and all related parts, which do not need the highest precision, are bonded by UV cured adhesives. Main drivers of this process technology, which is still evolving, are the availability of photoinitiators and LED as light sources close to the visible spectrum.In addition, the current system is capable of high processing velocities occurring in this area [78]. Today's UV-adhesion or glueing applications do only have a quality monitoring for mechanical stress and visible checks for errors in the final glued product.
The advantage of using the here developed sensor concept is twofold; first the conversion monitoring will lead to a constant curing of each glued product and second the laser beam is highly suitable for specific measurements on very small surfaces e.g. glue drops.

9.1.2　Circuit board solder resist (solder mask) coating

Circuit board coating has been realized by solvent based formulations for many years. Due to the reduction of solvents in any kind of application, circuit board coaters try to change their technique as well. Today, two component systems are used by most of the companies. This technique still has some problems concerning coating quality and adhesion to the actual circuit board.
The sensor technology momentarily does not allow a direct monitoring of the polymerization of the two component systems, because of the sensor design for acrylate systems. Yet, the main principle can be applied with ease and after characterising the chemical constituents in the formulation, the laser frequencies are adapted to a absorption band connected to the polymerization.
Furthermore, the applied coating thickness of 40 to $60\,\mu m$ and the high pigment concentrations are suited for a reliable conversion monitoring with this sensor concept as well as for the fast processing and the monitoring of specific areas on the board, where most care has to be taken.

9.1.3 Wood flooring and furniture coating

Wood flooring applications have been one of the first production areas, where UV coating techniques proved to be cost effective and predominant in surface hardness against solvent based coatings ([79] and [18]). In-line monitoring of the acrylate polymerization is often not implemented in the production because of economical reasons. Therefore the here presented sensor concept would increase the product reliability within the actual production process without additionally costs for extensive laboratory testing. The coating thickness of some tens to hundreds of micrometers is perfectly suited for the application of laser driven monitoring tools due to the high penetration depth in the polymer as shown in Section 6.3.

9.2 Advanced implementation scenarios

The issue of an absolute measurement, in this case the acrylate conversion, is always challenging and often not feasible. The main challenge is the complete interpretation of the acquired signal with respect to the degree of conversion. A possible solution is drafted in subsection 9.2.1. The final acrylate conversion is calculated in relation to the unpolymerized coating right at the same measurement spot on the product.
A conceivable ground-breaking evolution of the sensor system is presented in subsection 9.2.2. The proposed scenario is an expansion for gathering not only the conversion signal, but also information on coating thickness and coating application quality of complex three dimensional objects, for example metal food and beverage cans or automotive parts [80].

9.2.1 Comparison measurement with two sensor heads

The current sensor system might be expanded by using two sensor heads in combination with an optical multiplexer to measure the conversion of each sample point before and after the UV radiation exposure. Thus each data point is measured with respect to zero conversion cancelling out all other influencing effects, which might affect the conversion value. A absolute measurement of the conversion would be realized.
Figure 9.1 shows this set-up in combination with the UV-radiation source for a typical application scenario on a conveyor belt.

The set-up in Figure 9.1 is the most obvious expansion without changing curial parts like the sensor head and could be implemented in a short time

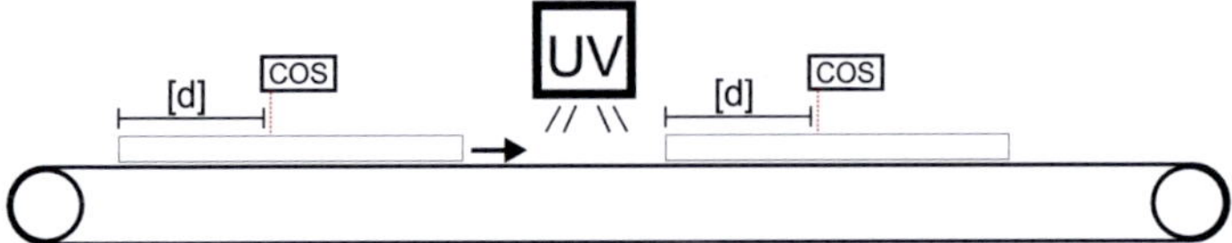

Figure 9.1: Outlook for a two sensors set-up also measuring the unpolymerized coating. Each measurement on the product is performed twice before and after UV-irradiation.

period.

First preliminaries are made by the IST Metz GmbH for applications in the printing industries.

9.2.2   Implementation as a scanner for three dimensional complex products

Additional development of the sensor system is most important for a further improved reliability and some advanced applications areas in future. A possible extension of the system is achievable by extending the system from static to scanning measurements. The advantages are numerous; complex coatings on three dimensional surfaces could be monitored in manifold ways, for example coating thickness before and after UV-curing or spray coating protection caps being at the right place, if labelled with IR-markers.

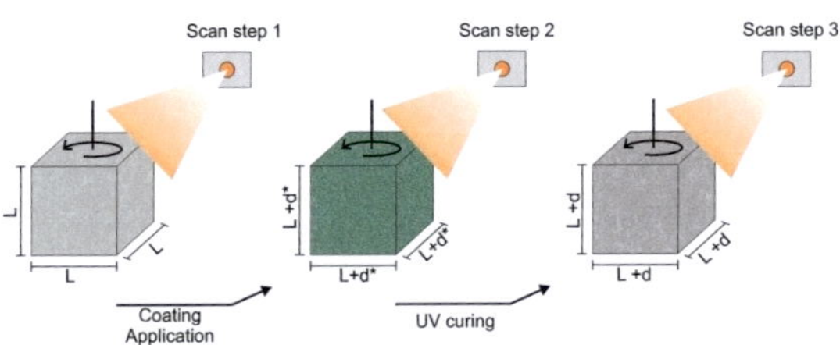

Figure 9.2: Outlook for a 3D scanner scenario on products like beverage cans or food packaging. The product is scanned by the sensor and several curial parameters like coating thickness, polymer conversion and surface finish are monitored.

The research complexity of this scenario must not be underestimated. The simple InGaAs-diode must be replaced by a detector array and the laser source must be adapted to a moveable scanner. But due to nowadays trend to highly integrated devices with multi-parameter acquisition, this kind of future sensor type might offer additional benefits for quality and thus as well for economical value.

Figure 9.2 shows the discussed application with a three-dimensional application scenario.

Chapter 10

Summary

A discrete near infrared laser spectroscopic measurement technique for monitoring the conversion (grade of polymerization) of UV-curing acrylates has been developed and it's applicability has been proved by measurements on various acrylic formulations. The main advantages compared to other methods like ATR-FTIR are the applicability of the system as an in-line monitoring tool and the acquisition of transflected laser signals from the coating used for the conversion measurement. The high energy density of the laser beams enables the system to reach the substrate of the coating and thus to gather an integral conversion value of the coating layer.

Several UV-cured acrylic formulations and conventional two-component curing formulations, which are used by the industry in various applications, have been examined in the far infrared with respect to their characteristic absorption differences before and after polymerization. The C-double bond cracking in the acrylate end group by the photoinitiator, launching the photopolymerization process, has been observed by absorption changes at $809\,cm^{-1}$ and $1406\,cm^{-1}$ in the far infrared spectrum ($600 - 1800\,cm^{-1}$). The correlated change in absorbance was proved to be observable in a second overtone oscillation of the C-H ligand in the near infrared at $1620\,nm$, as well.

For economical and technical reasons, the developed laser sensor system was designed to operate in the near infrared region at $1620\,nm$. An approximation model for the detection of the acrylic NIR spectra was successfully implemented. The detection method has been to measure the laser radiation at $1620\,nm$ transflected by the coating before and after polymerization.

Additionally to the DFB laser diode with the emission at $1620\,nm$ other discrete laser sources were implemented for normalization of the signal at $1620\,nm$.

The sensor system is able to detect the conversion values throughout the coating layer in a thickness range of $\approx 10\,\mu m$ to several millimetres for transparent or slightly pigmented coatings. High pigmented coatings with up to 10% of TiO_2 pigments could be analyzed in the range of $\approx 20\,\mu m$ to $\approx 200\,\mu m$. However, the absorbance of the detected second overtone at $1620\,nm$ is weak, compared to that of the fundamental modes in the far infrared and hence highly dependent on the initial amount of acrylic C-double bonds in the unpolymerized formulation. For this reason, a reliable detection is only possible, if the formulation is highly reactive and does not contain too many additives and other fillers.

The developed model formulation, a highly reactive transparent formulation consisting of TPGDA and Epoxy-acrylate, showed a relative signal difference (unpolymerized to polymerized) of 0.7% up to 6.3% with increasing coating layer thickness values from $16\,\mu m$ to $100\,\mu m$, respectively.

A validation of the sensor results by conventional ATR-FTIR and VIS/NIR measurements has proved the reliability of the detected conversion values. The initial conversion resolution of $\pm 20\%$ with the model formulation has been improved during the development process to a resolution of $\pm 10\%$ in the final prototype.

The laser signal transflected from the coating was examined in combination with different substrates, like metal or paper. Several angular scattering profiles were recorded and fitted by an analytical function. The best possible analytical description was found to be a combination of Gaussian and Lambert scattering. Metal substrates in combination with applied coatings show a Gaussian behaviour and small amounts of Lambert scattering. The pure Lambert scattering is only observable on paper substrates.

The Gaussian part of the scattered laser radiation was found to be the main direct reflection from the coating and the substrate surface. This part of the signal is only able to interact with the coating on it's rather short travel path of approximately twice the coating thickness, resulting in a only small detectable absorption change caused by the polymerization process.

The scattered radiation part, which produces a Lambertian characteristic, is better suited for the conversion detection, because of the longer travel distance in the coating compared to the Gaussian radiation part. This result is verified by a independent measurement; the experimentally determined optimal distance of the spatial filter to the substrate surface. The optimal distance of $50\,mm$ for the spatial filter, used in the sensor head, to the substrate has been found by observing the highest absorption difference between a unpolymerized and polymerized sample. This set-up does not deflect any radiation within an transflection angle of $-2°$ to $+2°$ to the detector. That Gaussian part is lost trough the pinhole mirror, whereas the Lambert part is reflected into the detector.

The developed discrete spectroscopy method for acrylates is highly suitable for quality control duties on sensible products, like food or medical packaging, which are produced in roll printing applications. The current prototype is not yet capable of detecting such thin coating layers in the range of some micrometers, due to a low signal to noise ratio, but it is already capable of measuring small spot sizes at high conveyor belt velocities. Further advances in DFB laser efficiency and detector sensitivity might be helpful in solving this issue. Next to that application field, the emerging UV-coating technologies for automotive applications need a reliable conversion monitoring, e.g. for corrosion protection coatings. Currently used FTIR methods are hardly suited to measure the conversion on bulky and complex products like transmission-bell housings. The discrete laser concept can be adapted to this task with reasonable costs and development effort.

Bibliography

[1] *Plastic separation of automotive waste by superfast near-infrared sensors*; Automotive Materials Recycle Bibliography; Argonne National Laboratory; 2002

[2] *NIR Spectroscopy as Powerful Tool for Process Control in UV Curing*; Conference Proceedings; RadTech Europe 2007; 2007

[3] Laub H.; *Irradiation Device*; Patent; 1940

[4] Directive 2004/42/EC of the European Parliament and of the council on the limitation of emissions of volatile organic compounds due to the use of organic solvents in certain paints and varnishes and vehicle refinishing products and amending Directive 1999/13/EC; *Official Journal of the European Union*; L 143:p. 87; 2004

[5] European Markets for Radiation Curable Coatings; *Tech. Rep. B919-01*; Frost and Sullivan; 2006

[6] Regulation (EC) No 1935/2004 on materials and articles intended to come into contact with food and repealing Directives 80/590/EEC and 89/109/EEC; *Official Journal of the European Union*; L 338:pp. 4–17; 2004

[7] Odian G.; *Principles of Polymerization*; Wiley & Sons; 4th edn.; 2004

[8] Glöckner P., Jung T. *et al.*; *Radiation Curing for Coatings and Printing Inks: Technical Basics, Applications and Trouble Shooting*; European Coatings Tech Files; Vincentz Network; 2008

[9] Fouassier J.P. and Rabek J.F., (eds.) *Radiation curing in polymer science and technology: Photoinitiating systems*; vol. 2; Elsevier Applied Science; 1993

[10] Photoinitiators for UV Curing; *Tech. rep.*; Ciba Specialty Chemicals; 2003

[11] Decker C.; The use of UV irradiation in polymerization; *Polymer International*; 45(2):pp. 133–141; 1998

[12] Studer K., Decker C. *et al.*; Overcoming oxygen inhibition in UV-curing of acrylate coatings by carbon dioxide inerting: Part II; *Progress in Organic Coatings*; 48(1):pp. 101–111; 2003

[13] Zhang Y., Kranbuehl D.E. *et al.*; Study of UV Cure Kinetics Resulting from a Changing Concentration of Mobile and Trapped Radicals; *Macromolecules*; 41(3):pp. 708–715; 2008

[14] Studer K., Decker C. *et al.*; Overcoming oxygen inhibition in UV-curing of acrylate coatings by carbon dioxide inerting, Part I; *Progress in Organic Coatings*; 48(1):pp. 92–100; 2003

[15] Pieke S.; *Experimentelle Untersuchungen zur effizienten Vernetzung von Oberflächenbeschichtungen mit UV-Strahlung*; Phd thesis; Karlsruher Institute of Technology; 2010

[16] Neo W.K. and Chan-Park M.B.; Application of a new model and measurement technique for dynamic shrinkage and conversion of multi-acrylates photopolymerized at different UV intensities; *Polymer*; 48(11):pp. 3337 – 3348; 2007

[17] DIN 5031-7: Optical radiation physics and illumination engineering; terms for wavebands; *Tech. rep.*; Deutsches Institut für Normung e.V.; 1982

[18] Bongiovanni R., Montefusco F. *et al.*; High performance UV-cured coatings for wood protection; *Progress in Organic Coatings*; 45(4):pp. 359–363; 2002

[19] Yang G.M., Peng S. *et al.*; UV LED Curing in Inkjet Printing Applications; *24th International Conference On Digital Printing Technologies, Technical Program and Proceedings*; pp. 535–537; 2008

[20] Bauer F., Decker U. *et al.*; UV curing and matting of acrylate nanocomposite coatings by 172 nm excimer irradiation; *Progress in Organic Coatings*; 64(4):pp. 474 – 481; 2009

[21] Oppenländer T.; Mercury-free sources of VUV/UV radiation: application of modern excimer lamps (excilamps) for water and air treatment; *Journal of Environmental Engineering and Science*; 6(3):pp. 253–264; 2007

[22] Wang D., Oppenländer T. *et al.*; Comparison of the Disinfection Effects of Vacuum-UV (VUV) and UV Light on Bacillus subtilis Spores in Aqueous Suspensions at 172, 222 and 254Â nm; *Photochemistry and Photobiology*; 86(1):pp. 176–181; 2010

[23] DIN EN 13523-11 Coil coated metals - Test methods - Part 11: Resistance to solvents (rubbing test); *Tech. rep.*; Deutsches Institut für Normung e. V.; 2010

[24] ASTM D740 - 05 Standard Specification for Methyl Ethyl Ketone; *Tech. rep.*; ASTM International; 2005

[25] DIN EN 15766, Packaging - Flexible aluminium tubes - Test methods to determine the polymerization of the internal coating with acetone;; *Tech. rep.*; Deutsches Institut für Normung e. V.; 2009

[26] ASTM D3359 - 09e2 Standard Test Methods for Measuring Adhesion by Tape Test; *Tech. rep.*; ASTM International; 2009

[27] ISO 1522:2006, Paints and varnishes – Pendulum damping test; *Tech. rep.*; International Organization for Standardization; 2006

[28] Wilson E.B., Decius J.C. *et al.*; *Molecular vibrations : the theory of infrared and Raman vibrational spectra*; McGraw-Hill, New York [u.a.]; 1955

[29] Zerbi G. and Del Zoppo M.; *Vibrational Spectra as a Probe of Structural Order/Disorder in Chain Molecules and Polymers, in Modern Polymer Spectroscopy*; Wiley-VCH Verlag GmbH, Weinheim, Germany; 2007

[30] Pavlyuchko A., Vasilyev E. *et al.*; Calculations of molecular ir spectra in the overtone and combination frequency regions; *Journal of Applied Spectroscopy*; 78(5):pp. 639–645; 2011-11-01

[31] Wheeler O.H.; Near Infrared Spectra Of Organic Compounds; *Chemical Reviews*; 59(4):pp. 629–666; 1959

[32] Fox J.J. and Martin A.E.; Investigations of Infra-Red Spectra. Determination of C-H Frequencies (3000 cm^{-1}) in Paraffins and Olefins, with Some Observations on "Polythenes"; *Proceedings of the Royal Society of London. Series A, Mathematical and Physical Sciences*; 175(961):pp. pp. 208–233; 1940

[33] Koirtyohann S.; A history of atomic absorption spectroscopy; *Spectrochimica Acta Part B: Atomic Spectroscopy*; 35(11-12):pp. 663 – 670; 1980

[34] Griffiths P.R. and De Haseth J.A.; *Fourier transform infrared spectrometry*; Wiley; 1986

[35] Mirabella F.M.; Internal Reflection Spectroscopy; *Applied Spectroscopy Reviews*; 21(1-2):pp. 45–178; 1985

[36] Harrick N.J. and du Pré F.K.; Effective Thickness of Bulk Materials and of Thin Films for Internal Reflection Spectroscopy; *Applied Optics*; 5(11):pp. 1739–1743; 1966

[37] Harrick N.J.; Electric Field Strengths at Totally Reflecting Interfaces; *J. Opt. Soc. Am.*; 55(7):pp. 851–856; 1965

[38] Technical Specifications for the LAMBDA 1050 UV/Vis/NIR and LAMBDA 950 UV/Vis/NIR Spectrophotometers; *Tech. rep.*; PerkinElmer Life and Analytical Sciences; 2007

[39] Applications and Use of Integrating Spheres with the LAMBDA 650 and 850 UV/Vis and LAMBDA 950 UV/Vis/NIR Spectrophotometers; *Tech. rep.*; PerkinElmer Life and Analytical Sciences; 2004

[40] Gather M.C., Koehnen A. *et al.*; White Organic Light-Emitting Diodes; *Advanced Materials*; 23(2):pp. 233–248; 2011

[41] Sandrowitz A., Cooke J. *et al.*; Flame emission spectroscopy for equivalence ratio monitoring; *Applied Spectroscopy*; 52(5):pp. 658–662; 1998

[42] Preliminary Technical Information Laromer® LR 9019; *Tech. rep.*; Performance Chemicals for Coatings, Plastics and Specialties, BASF Aktiengesellschaft; 2003

[43] Technical Information Laromer® TPGDA; *Tech. rep.*; Marketing Dispersionen, BASF Aktiengesellschaft; 67056 Ludwigshafen, Germany; 1996

[44] Scherzer T.; Depth Profiling of the Degree of Cure during the Photopolymerization of Acrylates Studied by Real-Time FT-IR Attenuated Total Reflection Spectroscopy; *Applied Spectroscopy*; 56:pp. 1403–1412(10); 2002

[45] Yang D.B.; Kinetic studies of photopolymerization using real time FT-IR spectroscopy; *Journal of Polymer Science Part A: Polymer Chemistry*; 31(1):pp. 199–208; 1993

[46] Hong B.T., Shin K.S. *et al.*; Ultraviolet-curing behavior of an epoxy acrylate resin system; *Journal of Applied Polymer Science*; 98(3):pp. 1180–1185; 2005

[47] Decker C.; Kinetic study of light-induced polymerization by real-time UV and IR spectroscopy; *Journal of Polymer Science Part A: Polymer Chemistry*; 30(5):pp. 913–928; 1992

[48] Bajpai M., Shukla V. *et al.*; Film performance and UV curing of epoxy acrylate resins; *Progress in Organic Coatings*; 44(4):pp. 271 – 278; 2002

[49] Beuermann S., Buback M. *et al.*; Kinetics of free radical solution polymerization of methyl methacrylate over an extended conversion range; *Macromolecular Chemistry and Physics*; 196(8):pp. 2493–2516; 1995

[50] Scherzer T., Mehnert R. *et al.*; On-line monitoring of the acrylate conversion in UV photopolymerization by near-infrared reflection spectroscopy; *Macromolecular Symposia*; 205(1):pp. 151–162; 2004

[51] Aldridge P.K., Kelly J.J. *et al.*; Noninvasive monitoring of bulk polymerization using short-wavelength near-infrared spectroscopy; *Analytical Chemistry*; 65(24):pp. 3581–3585; 1993

[52] Urban M.W. and Provder T., (eds.) *Multidimensional Spectroscopy of Polymers*; American Chemical Society, Washington, DC; 1995

[53] Weyer L.G.; Near-Infrared Spectroscopy of Organic Substances; *Applied Spectroscopy Reviews*; 21(1):pp. 1–43; 1985

[54] Heymann K., Mirschel G. *et al.*; Monitoring of the Thickness of Ultraviolet-Cured Pigmented Coatings and Printed Layers by Near-Infrared Spectroscopy; *Applied Spectroscopy*; 64(4):pp. 419–424; 2010

[55] Egan W.G., Hilgeman T. *et al.*; Determination of Absorption and Scattering Coefficients for Nonhomogeneous Media. 2: Experiment; *Appl. Opt.*; 12(8):pp. 1816–1823; 1973

[56] Jafarzadeh S., Adhikari A. *et al.*; Study of PANI-MeSA conducting polymer dispersed in UV-curing polyester acrylate on galvanized steel as corrosion protection coating RID C-3934-2009; *Progress In Organic Coatings*; 70(2-3):pp. 108–115; 2011

[57] Sze Simon Min ; Ng K.K.; *Physics of semiconductor devices*; Wiley, New York, NY; 3. ed. edn.; 2007

[58] Kogelnik H. and Shank C.V.; Coupled Wave Theory of Distributed Feedback Lasers; *Journal of Applied Physics*; 43(5):pp. 2327–2335; 1972

[59] Sato T., Mitsuhara M. *et al.*; InAs Quantum-well Distributed Feedback Lasers Emitting at 2.3 μm for Gas Sensing Applications; *NTT Technical Review: Special Feature: Light Source Technologies for Sensing Applications*; 7(1):pp. 1–7; 2009

[60] Koeth J.; Quantum-dot lasers provide high performance near 1.15 microns; *SPIE Newsroom*; 2008

[61] Zeller W., Naehle L. *et al.*; DFB Lasers Between 760 nm and 16 μm for Sensing Applications; *Sensors*; 10(4):pp. 2492–2510; 2010

[62] Lindner D.; Messmethoden zur Kalibrierung von Lichtstärke Normallampen und Leuchtdichte Normalen sowie deren Eigenschaften; *18. Gemeinschaftstagung der Lichttechnischen Gesellschaften der Deutschlands, Österreichs, der Schweiz und der Niederlande*; 2008

[63] Model 7265, DSP Lock-in Amplifier, 190284-A-MNL-C; *Instruction manual*; Ametek Advanced Measurement Technology Inc., Signal Recovery; 2002

[64] Decker C. and Moussa K.; A new method for monitoring ultra-fast photopoly- merizations by real-time infra-red (RTIR) spectroscopy; *Die Makro- molekulare Chemie*; 189(10):pp. 2381–2394; 1988

[65] Decker C., Nguyen Thi Viet T. *et al.*; UV-radiation curing of acrylate/epoxide systems; *Polymer*; 42(13):pp. 5531–5541; 2001

[66] Bohren C.F.; *Absorption and scattering of light by small particles*; Wiley; 1983

[67] Fabbri F., Franceschini M.A. *et al.*; Characterization of Spatial and Tempo- ral Variations in the Optical Properties of Tissuelike Media with Diffuse Reflectance Imaging; *Applied Optics*; 42(16):pp. 3063–3072; 2003

[68] Nakano T., Shimada S. *et al.*; Transient 2D IR Correlation Spectroscopy of the Photopolymerization of Acrylic and Epoxy Monomers; *Applied Spec- troscopy*; 47(9):pp. 1337–1342; 1993

[69] Decker C. and Moussa K.; Kinetic study of the cationic photopolymerization of epoxy monomers; *Journal of Polymer Science Part A: Polymer Chem- istry*; 28(12):pp. 3429–3443; 1990

[70] Decker C.; Photoinitiated crosslinking polymerisation; *Progress in Polymer Science*; 21(4):pp. 593–650; 1996

[71] Mijovic J., Andjeli S. *et al.*; In situ real-time monitoring of epoxy/amine kinetics by remote near infrared spectroscopy; *Polymers for Advanced Technologies*; 7(1):pp. 1–16; 1998

[72] de Souza F.G., Anzai T.K. *et al.*; In situ determination of aniline polymer- ization kinetics through near-infrared spectroscopy; *Journal of Applied Polymer Science*; 112(1):pp. 157–162; 2009

[73] Leclercq L., Maschke U. *et al.*; Light scattering from acrylate-based polymer dispersed liquid crystals: theoretical considerations and experimental ex- amples.; *Liquid Crystals*; 26(3):pp. 415 – 425; 1999

[74] Fixman M.; Scattering from Polydisperse Melts; *Macromolecules*; 37(22):pp. 8441–8456; 2004

[75] Bruker Optics; *Alpha, FTIR-Spectrometer Manual*; 2007

[76] Scherzer T., Muller S.and Mehnert R. *et al.*; On-line monitoring of the con- version in photopolymerized acrylate coatings on polymer foils using NIR spectroscopy; *Polymer*; 46(18):pp. 7072–7081; 2005

[77] Colthup N.B., Daly L.H. *et al.*; *Introduction to infrared and raman spectroscopy*; Academic Press, New York, NY [u.a.]; 1964

[78] Bachmann A.; Advances in light curing adhesives; *Spie Conference, San Diego*; 444-20; 2001

[79] UV-Curable Wood Furniture Coatings Case Studies; *Tech. rep.*; United States Environmental Protection Agency (EPA); 2001

[80] Sturm Maschinenbau GmbH and Haimerl C.; UV-Lackieranlage für Lkw-Achsen; *Journal für Oberflächentechnik und Beschichtung*; 3:pp. 2–5; 2010

List of Figures

List of Tables

Acknowledgements

First of all I want to thank Prof. Wolfgang Heering for his excellent support and helpful advises during this work and to have the opportunity to conduct this work at the Light Technology Institute. Prof. Heerings advises and technical expertise were always important and contributed significantly to the success of my work during the last four years.

Second I want to thank Prof. Wilhelm Stork for his interest in my work and his readiness for being co-examiner of my thesis.

This thesis would not have been possible without the fruitful discussions and outstanding help of Dr. Oliver Treichel and Elmar Feuerbacher. It was always a pleasure to work with you and I hope we can realise some other projects in near future.

I have to thank all my current and former co-workers in the Light- and Plasma group and the complete staff at the LTI for their support and ongoing scientific contributions. Especially Dr. Rainer Kling, Christoph Kaiser, Mohan Celal Ögün, Michel Meisser, Karsten Hähre and Stefan Pieke.

I have to thank the companies IST Metz GmbH, Robert Bosch GmbH and QIAGEN Lake Constance for their support with material and laboratory equipment. Special thanks to Robert Kretzschmar, Michael Stengle and Dr. Josef Drexler.

Thanks to the professional team of the LTI work shop for the uncountable customized equipment and our administration staff for always defending the project funds. This work is based upon a project funded by the Bundesministerium für Bildung und Forschung (BMBF), I want to thank the BMBF for the financial support and especially I want to thank Dr. Alexander Lucumi for the project supervision.

Finally I would like to thank Marit for her outstanding support during the four years working on this thesis.

Curriculum Vitae Mathias Bach

Personal details

Date of birth   27. August 1981 in Mutlangen, Germany

Nationality   German

Education

1992 - 2001   Rosenstein Gymnasium, Heubach

2002 - 2008   Master in Geophysics,
University Karlsruhe (TH)

2008 - 2012   PhD in electrical engineering,
Karlsruhe Institute of Technology, KIT

Work experince

2001 - 2002   Work and Travel Australia

2005 - 2006   Head of geoelectrical and sesmic field campaigns,
University Karlsruhe (TH)

2008 - 2012   Scientist, Member of the
Light- and Plasmatechnologies Group,
Light Technology Institute, KIT